Exam Prep Course

The Best Boiler Operator Exam Prep Course

The Best Boiler Operator

Exam Prep Course
Foreword:

This manual will serve as your refresher course for the study of boiler operation and principles. It is called Your Boiler Operator Exam Pocket Tutor because it is told through simple explanatory language. This book and the accompanying audio version will serve the reader and listener in more of a tutoring capacity.

Our approach taught here in Pocket Tutor is a proven method that has proven to be immensely successful in helping students get their boiler operator license. Our goal is to help readers gain the necessary confidence in their competence relative to Boiler Operation that they will successfully pass their boiler operator license examination in the jurisdiction they reside. Many people embarking on the study of boiler operation often take on too much study material and get overwhelmed from all the available literature. Or students try to learn from Cliff Notes from friends or fellow students that may or may not have been successful in earning their boiler operator license.

This book takes a tutor approach rather than an instructor approach. Our approach will have the student fully prepared to have talk about boilers, boiler auxiliaries and accessories, their operation and testing in a conversation manner and tone. This manual will demystify boilers and their operation.

Copyrights Reserved ©2019

No part of this book can be transmitted or reproduced in any form including print, electronic, photocopying, scanning, mechanical or recording without prior written permission from the author.

While the author has taken utmost efforts to ensure the accuracy of the written content, all readers are advised to follow the information mentioned herein at their own risk. The author cannot be

THE BEST BOILER OPERATOR

held responsible for any personal or commercial damage caused by misinterpretation of information. All readers are encouraged to seek professional advice when needed.

This book has been written for information purposes only. Every effort has been made to make this book as complete and accurate as possible. However, there may be mistakes in typography or content. Also, this book provides information only up to the publishing date. Therefore, this book should be used as a guide - not as the ultimate source.

The purpose of this book is to educate. The author and the publisher do not warrant that the information contained in this book is fully complete and shall not be responsible for any errors or omissions. The author and publisher shall have neither liability nor responsibility to any person or entity concerning any loss or damage caused or alleged to be caused directly or indirectly by this book.

Exam Prep Course
Chapter 1. Boiler Industry

What is a Boiler Plant?

Ans. A plant consisting of one or more boilers in battery or connected to a common header or steam outlet used to generate steam for heating, process or mechanical work in a prime mover.

What is a heating plant?

Ans. A boiler plant consisting of one or more boilers in battery connected to a common header or steam outlet used to generate steam for heating conditioned spaces. Depending on the distances the steam must travel to the condition space will determine the pressure the boiler must generate. The steam once it leaves the boiler is then directed and handled by other auxiliaries and accessories before being discharged through heat transfer surfaces which give off the heat energy in the steam to the conditioned space before returning to condensate and returning to the boiler via condensate return pumps, condensate tank and boiler feed pumps.

The heat energy produced by a boiler in a heating plant is often recycled and reused in other processes to reduce fuel consumption in the boiler and thus reduce overall costs to operate and maintain the boiler. The steam leaving the boiler may also be reduced through Pressure Reducing Stations or Pressure Reducing Valves (PRVs) before the steam enters heat exchanging equipment. If the heat exchanging equipment (radiators or coils) are relatively in close proximity to the boiler a PRV is not required as the steam pressure entering the coil is sufficient to heat the space and not do damage to the coil as it gives off its heat energy and return to condensate.

THE BEST BOILER OPERATOR

Heat exchange equipment in a heating plant must be of sufficient size to give off enough British Thermal Units (BTUs) to heat the space to the design temperature as well as allow the steam to flow freely through the coil with minimum condensate logging which reduces the flow of steam and reduces the capacity of the coil because the steam is now heating condensate in the coil and not giving off heat to the coil to heat the area or medium surrounding.

What is a Power Plant?

Ans. A boiler plant consisting of one or more boilers in battery connected to a common header or steam outlet used to generate steam for delivering to a prime mover such as a turbine which is directly or indirectly connected to an electrical generator. The turbine turns from the steam being directed through its nozzles and in turn the generator connected to it turns. This turning of the generator through its own processes produce voltage which is transmitted out for use.

Because Power Plants use very large steam boilers and complex systems the steam generated by the boiler is not simply used to turn a turbine. It is also used to demineralize water for the boiler's use. The steam is used to heat various spaces inside the plant and for other processes that the plant is designed.

What is a low pressure and a high pressure boiler?

Ans. The classification of low and high pressure boilers are determined by the American Society of Mechanical Engineers (**ASME**). Low Pressure boilers are any operating under 15 pounds per square inch (psi). Any boiler operating 15 psi and higher is considered a high pressure boiler.

Exam Prep Course

Low pressure boilers are typically used in low grade commercial and industrial heating plants. Schools and hospitals that do not use steam for process or power will use a low pressure boiler or a steam generator to provide the heat energy it needs to condition a space.

Compared to low pressure boilers, high pressure boilers are much more vast in the types since any boiler operating over 15 psi is considered a high pressure boiler.

How are boilers used in laundries?

Boilers are used laundries generate steam to operate presses and other mechanical devices used to clean and launder clothes. The steam is high pressure and if not needed reduced by a PRV for the particular purpose.

How are boilers used in process plants?

Boilers used in process plants generate steam to finish processes like beer and wine; operate prime movers or equipment needed in the manufacturing process. Steam in this case is not used for its heat energy per se but for the applied force or pressure it can exert upon a surface to get it do work.

What is a Power Generating Plant?

Boilers used in power generating plants produce steam by converting the chemical energy in the fuel into heat energy in the steam. The heat energy and the pressure generated sends the steam long distances so that it can do work in the turbine which is connected to a generator that produces electricity for transmission and usage.

In the power generating plant no different than any other plant the boiler converts the stored chemical energy in the fuel into heat energy in the steam. The boiler's ability to do

The Best Boiler Operator

this in an efficient and safe manner is what determines how effective the plant is overall. Power plants want to generate the most electricity in watts for the cheapest cost which is heavily driven by the fuel it takes to produce a pound of steam. That cost plus the cost from all the accessories associated in the boiler plant and the labor give an idea of the total cost.

Exam Prep Course
Chapter 2. Boiler Purpose

What is a boiler?

Ans. An energy conversion device that converts the potential stored chemical energy in the fuel combusted in the furnace and converts into heat energy that is transferred to the water through the heating surfaces to become steam.

A **boiler** or **steam generator** is a device used to create steam by applying heat energy to water. Although the definitions are somewhat flexible, it can be said that older steam generators were commonly termed boilers and worked at low to medium pressure (7–2,000 kPa or 1–290 psi) but, at pressures above this, it is more usual to speak of a *steam generator*.

A boiler or steam generator is used wherever a source of steam is required. The form and size depends on the application: mobile steam engines such as steam locomotives, portable engines and steam-powered road vehicles typically use a smaller boiler that forms an integral part of the vehicle; stationary steam engines, industrial installations and power stations will usually have a larger separate steam generating facility connected to the point-of-use by piping. A notable exception is the steam-powered fireless locomotive, where separately-generated steam is transferred to a receiver (tank) on the locomotive.

How is the term defined and how do you make sense of it in a common way to be able to comprehend its function and its role in a either a heating or power plant?

Ans. Well we know that a boiler by scientific term is an enclosed vessel when operating under pressure converts water into steam. That definition has been accepted and as good

The Best Boiler Operator

as any provided for almost a century.

However that response treats the boiler separate or above the rest of the plant when actually it is just a part of the plant's purpose or operations.

What is a boiler's purpose?

Ans. A boiler's sole and only job is to convert the chemical energy in the fuel to heat energy in the steam to do work in the designed application. It does not do anything beyond that. The boiler is not concerned with the steam beyond that.

Therefore the operator must not over complicate or emphasize the boiler's importance. It may be the lead singer of the power plant band but it is still just a member of the group. Without proper understanding of the associated auxiliaries and accessories working together then the boiler is useless.

How are boilers designed?

Ans. Each boiler type is designed a certain way to allow the greatest absorption of heat from the fuel burned in its furnace area transferring that heat to the water inside to create the driest steam leaving the steam drum or outlet. That is it. That is all. Understand that every component inside the boiler proper (inside) is designed to do those three things:

1. Absorb the greatest amount of heat from the fuel in the furnace in the safest manner
2. Transfer that heat with minimal loss to the water inside the boiler
3. Deliver the highest quality of steam for the pressure and application of the system
4. Help support the boiler structure and allow for the expansion and contraction of the boiler as it is heated and cooled

Exam Prep Course

So now let's add what you just learned back to the old definition to give you a deeper understanding of a boiler's purpose. A boiler is a closed vessel in which water is transformed into steam through the application of heat is the old definition. We recommend a new definition:

A boiler is an energy conversion device that converts the potential stored chemical energy in the fuel combusted in the furnace and converts into heat energy that is transferred to the water through the heating surfaces to become steam. A boiler could not be a boiler if it was not an enclosed vessel. Saying it is a pressure vessel is severely a fraction of what a boiler is or what it does.

The Best Boiler Operator
Chapter 3. Boiler Types & Classifications.

Boilers can be classified in many ways. Understanding the various boiler types and the way in which they can be classified will go a long way in comprehending how they accomplish their purpose. There are two types of boilers: **Fire Tube and Water Tube**. Boilers can be classified in numerous ways such as the axis of the shell: **horizontal or vertical**. Boilers can be classified by the position of the furnace whether it is **internally or externally fired**.

Fire tube boilers are boilers in which the products of combustion or hot gases from combustion travel from the furnace area through the boiler tubes with the boiler water outside. The tubes are connected to tube headers on both the front and rear of the boiler with cylindrical drum make up the boiler shell. Think of a pop can laying on its side (horizontal) or standing up (vertical) filled with tubes. The flat ends of the pop can make up the tube sheet that hold the tubes. Gases passes through the tubes with water surrounding.

What are some of the advantages of using a firetube boiler?

Ans. Firetube boilers contain a large amount of water due to its design with the water being on the around the tubes and filling a great deal of the shell. Because of this design firetube boilers can respond to load changes with minimal steam pressure drop.

Are firetube boilers fast steamers?

Ans. Yes! Firetube boilers are fast steamers. This is because of the large heating surface provided by all the tubes, the tube sheet and in cases like the Scotch Marine Boiler where the furnace is located inside the shell and extends back 60% the length of the drum creates another large heating surface to quickly generate steam.

What is the circulation of hot gases and water inside a firetube boiler?

Ans. Gases travel from the furnace either by being pushed through the boiler from a forced draft fan creating a positive pressure. The fan must be large enough or have enough capacity to overcome all of the obstacles in ways to create and maintain the proper draft pressure to force the products of combustion through the tubes and out of the stack.

As the water is heated is become lighter in density or weight. This lighter weight is also assisted by the heavier cooler water falling inside the shell and as it falls pushes or assist the lighter water rise. As the hotter water rises and cooler water falls it creates the circulation within the shell. The lighter hotter water rises until it breaks the water surface and into the steam space and steam baffle or steam separator before exiting the boiler shell.

What are the types of firetube boilers?

Ans. The main firetube boiler types are some variation of either a dryback or wetback boiler, firebox, vertical and H.R.T.

What is meant by a dryback firetube boiler?

Ans. The dryback is exactly what the term means. The brick lined furnace external to the boiler vessel directs the hot gases from the furnace to the tubes. The melting point of the brick is higher than the temperature of the hot gases and as long as the fire from combustion is not directly hitting the brick work it remains safe and capable of protecting the boiler shell from the direct heat.

The Best Boiler Operator

CUT AWAY VIEW

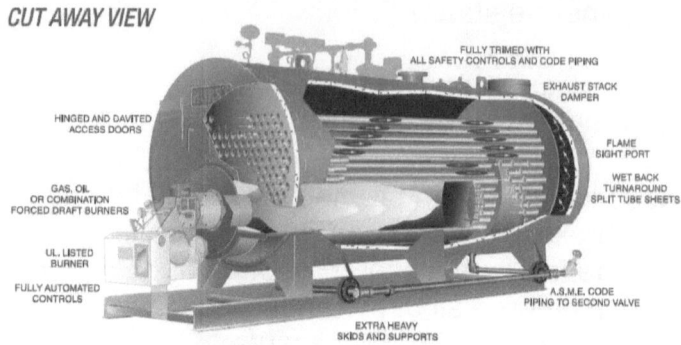

What is meant by a wetback firetube boiler?

Ans. The furnace directing the hot gases is totally surrounded by water creating a large heating surface and allowing the boiler to generate steam faster.

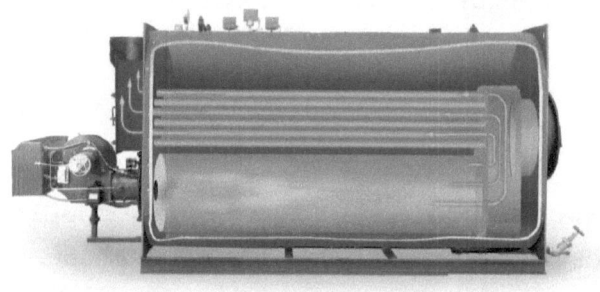

What are the drawbacks of a wetback boiler?

Ans. The weight of the furnace suspended and attached to the tubesheet is extremely heavy and could cause leakage around the seal of the furnace.

What is a firebox boiler?

Ans. a horizontal return firetube boiler that has a round top and flat sides. It is internally fired meaning the furnace is inside the boiler setting but it is not inside the cylindrical shell. This boiler has water legs that extend down the sides and increase the heating surface. You find these type of boilers in school buildings primarily in the Northeast and Midwest.

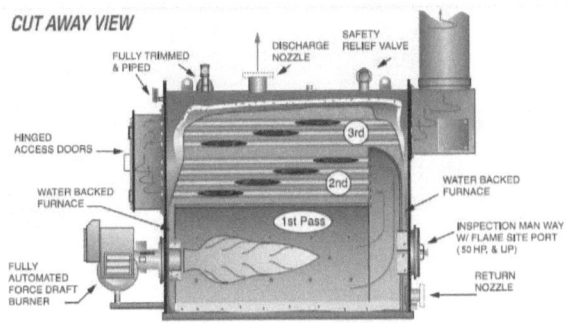

What is a H.R.T. boiler?

Ans. In a H.R.T., the cylindrical shell of the boiler containing the tubes and tube sheets is suspended over the furnace. The hot gases are directed from the front of the boiler over a bridge wall to the back tube sheet and a baffle directs them back to the front tube sheet where another baffle directs back to the rear where they discharge out the stack. The gases making multiple passes in the tubes are allowed to give up more heat to the water thus making it more efficient as it is able to generate more steam per pound of fuel used.

The drawbacks of the HRT is the entire cylinder is suspended. The weight of all this steel with the tubes and water becomes a heavy burden for the suspension slings. The constant heating and cooling of the metal shell also produces great stress on the steel beams holding the cylindrical shell in place.

THE BEST BOILER OPERATOR

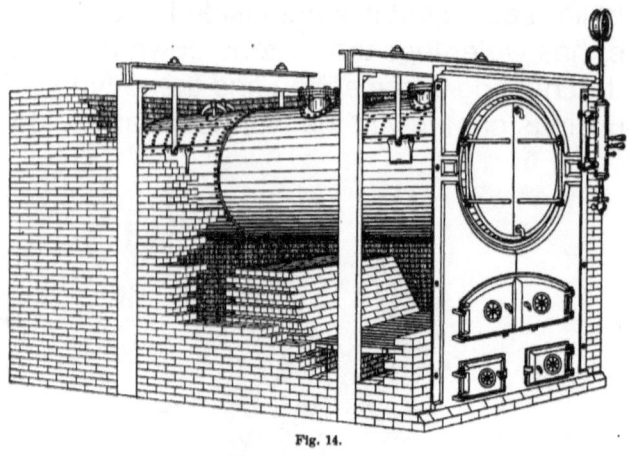

Fig. 14.

What is a water tube boiler?

Ans. In a water tube boiler, heat from the flames and gases of combustion is transferred to the water passing through the tubes. Baffles direct the gases of combustion for multiple passes to obtain maximum efficiency. Firetube boilers were limited by the size they would need to be to reach the higher pressures the boiler industry demanded. By moving the water in the tubes and having the gases on the outside allowed for boiler operating pressures to increase. The individual tube circuits allow for a greater heat transfer and larger structure than its firetube counterpart.

How are water tube boilers classified?

Ans. Industrial and Commercial

What is an industrial water tube boiler?

Multiple drum boiler. A steam (top) drum and a mud (bottom) drum. Used in process applications. Gases of combustion pass around the tubes transferring heat through to the tubes of water. Again as water is heated it rises and cooler falls creating a natural circulation. Water that enters the boiler enters in the coolest section and falls as riser tubes are located in the area where the greatest heat absorption occurs. This aids in circulation.

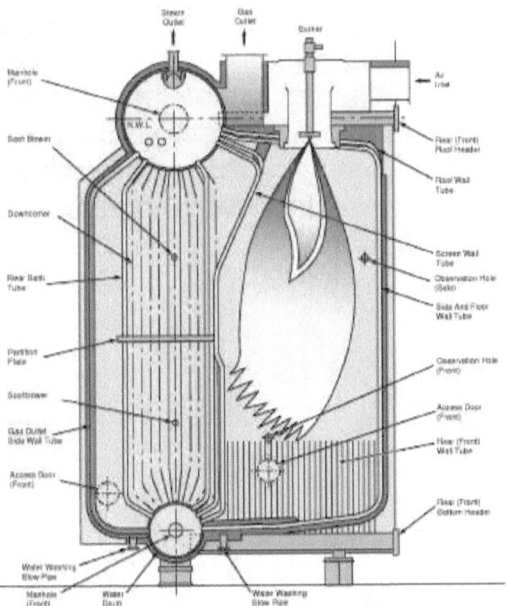

What is a problem that could occur with this boiler arrangement with boiler water circulation?

Ans. The gases of combustion are directed through the use of brick baffles. The baffles ensure that the gas flow in a manner that allows for the greatest amount of heat absorption. If the baffles are broken or compromised then the hot gases will not flow in the designed path. This will disrupt circulation as the downer tubes will receive more heat than they are supposed to and begin to act like risers. The natural circulation is what helps keep the tubes cool enough to withstand the high temperatures from the higher firing rate of the furnace. If not corrected the boiler front risers will be damaged as it will not be able to take heat away fast enough to stay under its melting point.

The Best Boiler Operator

What is a commercial boiler?

Ans. Is a watertube boiler used in commercial or medium-sized applications. These boilers are either bent tube or straight tube boilers. Whether the tube is bent is depending on the design and how the steam (top) drum and the mud (bottom) drum are connected. Bent tube boilers such as the D and O type allow for great combustion space and increased heating surface with the tubes surrounding the combustion chamber.

Exam Prep Course
Chapter 4: Boiler Safety

The most important aspect to operating boilers and their related components is the ability to do it safely. The failure to operate safely and or maintain the equipment in a safe manner poses imminent physical threat to the operator, building occupants and in some cases the surrounding community.

A boiler explosion happens more often than people think. In 2019, there was a boiler explosion in St. Louis that killed three. The explosion was caught on tape. https://abcnews.go.com/GMA/video/killed-boiler-explosion-caught-camera-46562554. This statement is not to lay blame for the explosion on anyone other than to emphasize that despite all the safety procedures and protocols that if not properly maintained or followed a boiler and its accessories pose a severe safety threat if ignored.

Boiler explosions occur from primarily two ways: shell explosion or combustion chamber explosion. A shell explosion is either from a sudden drop in pressure without a corresponding drop in temperature. The large amount of water that takes less space in the shell suddenly expands 1600 its size as it flashes into steam. The boiler shell cannot handle that sudden release of energy which is akin to a bomb going off. The other shell explosion results from the boiler's pressure rising above the shell strength and the safety valve located atop the highest part of the steam space fails to relieve all of

The Best Boiler Operator

the steam that boiler generates preventing it from exceeding **6%** of its ***Maximum Allowable Working Pressure*** (MAWP).

What is a safety valve?

Ans. A safety valve is an automatic, full-opening, pop-action valve, opened by overpressure in a boiler. The safety valve has a compression spring which holds the valve down on its seat that keeps it closed. This spring is opposing boiler pressure at all times. In fact the pressure the spring exerts downward (spring pressure) is being applied at all times.

It is only when the boiler pressure overcomes the spring pressure during an overpressure situation while the boiler is operating that spring begins to lift of its seat. Once it lifts off its seat, steam from the boiler is exposed to a larger area inside the valve called the huddling chamber. When the steam from the boiler reaches the huddling chamber its opposing force against spring pressure is suddenly larger and the valve pops open. The boiler pressure with the help of the huddling chamber keeps the valve open long enough to relieve the pressure that has exceeded the safety valve's spring set pressure.

Exam Prep Course

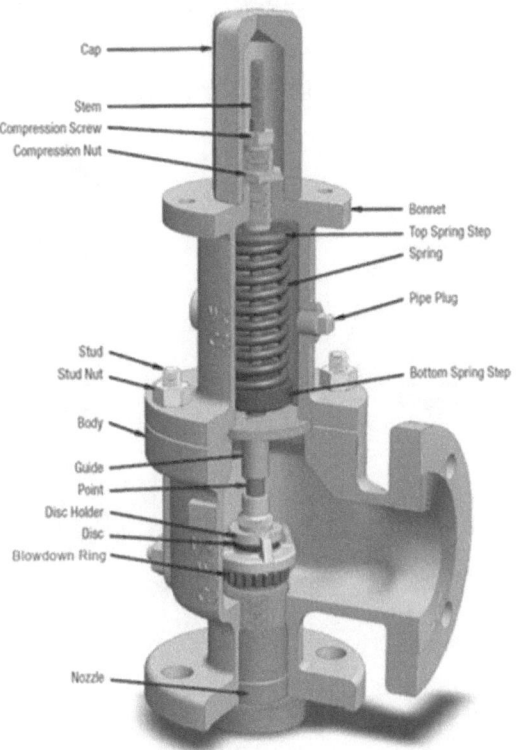

What is the difference between the popping pressure and set pressure of the safety valve?

Ans. Safety valves are set to pop at a predetermined pressure. However, due to time and errors in setting at the factory they are built the valve may not pop at the set pressure. The popping pressure is the pressure that the valve actually pops. The set pressure is the pressure listed on the safety valve data plate that the valve actually lifted during testing conditions in the factory.

What is the huddling chamber?

Ans. It is the increased area within the safety valve's body that the steam is exposed to once the steam pressure begins to overcome the safety valve spring pressure. The huddling chamber is what gives the safety valve its popping action when it opens rather than a slow lift of its seat like a relief valve found on water heaters or hydronic boilers.

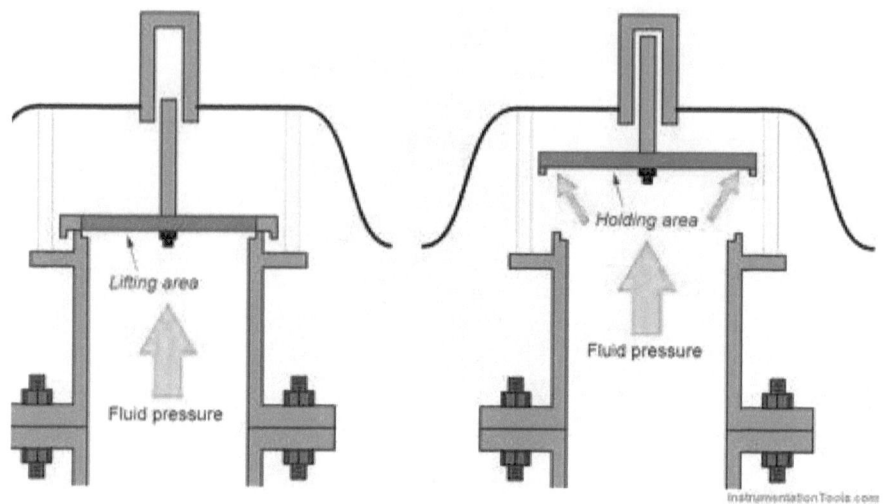

What is blowback?

Ans. Blow back is the difference in pressure from when the valve pops and when the valve reseats slamming shut from the spring pressure of the valve. Blowback prevents "chattering" of a safety valve.

Chattering is the rapid opening and closing of a safety valve. With blowback you ensure that enough steam has been relieved and the spring pressure of the safety valve is back in control to slam the valve shut. Too much blowback and you will waste steam out your outlet and too little will add unnecessary wear and tear on the valve from repeated opening and closing of the safety valve.

If the valve is used or ever actuated the Operator in Charge should test the valve again and note the popping and seating pressure as well as the blowback in pounds per square inch.

What is a high limit control?

Ans. A pressure control that senses steam pressure rising above a set point and shuts the burner down to stop the input of heat and thus the production of steam. The high limit control is a backup for the operating control which is the primary control for managing boiler operating pressure. High limit controls are required to have a manual reset so the operator must investigate and know that there was an overpressure situation that caused the boiler high limit to activate. If the high limit fails after the operating limit fails then your safety valve must prevent the boiler from exceeding 6% above its Maximum Allowable Working Pressure.

What is a low-water cutoff?

Ans. A boiler accessory that shuts off the burner in the event of a low water condition. The LWCO activates when the tricocks have sounded and after your feedwater regulator has failed to match the water needed with the steam leaving to keep a constant water level in the boiler.

The Best Boiler Operator

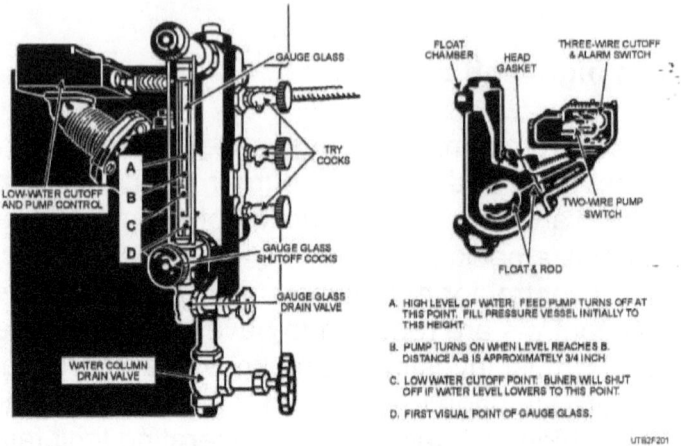

Where is the LWCO located?

Ans. Slightly below the Normal Operating Water Level. The top line connects to the highest part of the steam side of the boiler. The bottom line connects to the water side well below the NOWL. The chamber of the float on the LWCO has a blowdown, drain line and valve which should be activated daily or every shift depending on the boiler's operating capacity or pounds of steam generated.

How do you perform an evaporation test?

Ans. An evaporation test is performed to test the operation of the LWCO. To perform the operator shuts off the feedwater line to the boiler and keeps the boiler running producing steam. As the steam is generated and no water is replenished the water level drops and if operating properly the LWCO should send a signal to the burner to shut-down before the water level reaches an extraordinary low level. The LWCO operates when the feedwater regulator fails to do its job. The feedwater regulator first job is to maintain a constant water level in the boiler. If it is compromised then the boiler is at risk to damage tubes from overheating. The LWCO protects the boiler by shutting down the burner in a low water situation.

What is a flame safeguards?

Ans. It is a device that monitors the presence and condition of a flame during start-up of a boiler during the combustion process. If a boiler during its start-up misses a step in the combustion process the flame safeguard will either shut the flow of fuel to the furnace off thus stopping combustion or it will prevent the flow of fuel to the furnace if conditions are not correct.

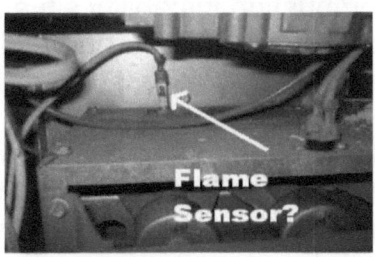

All flames, whether they are from burning wood, oil, gas, coal, or even a candle, produce heat and light (radiant energy), and during the burning process, ions are formed. The only difference between the various flames is in the magnitude of each of these quantities.

Some flames are blue, some are yellow, and some produce more heat than others do, but basically they are all the same. For instance, the atomized oil burner produces a characteristic yellow flame, but at the same time it is also radiating light in the infrared region and the ultraviolet region. You cannot see ultraviolet or infrared light, but the warmth you feel on your face is a result of the infrared radiation.

The ultraviolet radiation can only be detected by an ultraviolet sensitive photocell. The pale blue flame of a gas burner is familiar to us, and it is apparent how the visible light is present in both this and the oil flame. Both flames differ in brightness as seen by the eye as well as in color and in the amount of infrared and ultraviolet radiation they produce.

What are the types of flame safeguards?

Ans. Flame scanners, flame rods and photocells.

How does a flame scanner work and how do you test a flame scanner?

Ans. Primary target of the flame scanner is to alert operators about combustion instability situations potentially leading to the accumulation of unburned fuel and potential hazard for plant personnel.

The radiation from a flame in the visible region is easy to visualize because it is visible to the eye. Whenever a human eye can see a flame, it is radiating visible light. The infrared radiation from an oil flame is very similar in pattern to the visible radiation, except that it will be present slightly beyond the flame envelope. In a gas flame, the infrared radiation will be strongest in the last two-thirds of the flame. The area of the flame that radiates ultraviolet is much more limited than either the visible or infrared areas.

Practically all the ultraviolet radiation comes from the first one-third of the flame for all fuels that are premixed with air. For atmospheric flames, such as a candle or raw gas burning at the end of a pipe, the area of the flame radiating ultraviolet will be the outside surface of the flame.

The radiation from a flame constantly varies in amplitude at all times. That is, it is momentarily changing from brighter to dimmer to brighter and so on. We have all seen how some flames flicker at a low frequency, but even those flames that seem to be perfectly steady to the eye flicker.

A photoelectric detector can pick up this flicker. The rate of flicker is usually too fast to be seen by the eye. The frequency of modulation depends to some extent on the type of flame, but it has been found by a large number of measurements that a substantial amount of all flames flicker with rates ranging from 10 to 50 Hz.

The most familiar property of a flame is, of course, the high temperature of the gases within the flame and in the gas products of the flame. The maximum temperature of the flame depends on the fuel being burned and the availability of oxygen.

Flame failure control can utilize another characteristic of flames; the light transmission of the flame's combustion products. When a flame is burning without enough air for clean burning, there will be soot (free carbon) produced which causes a smoky flame and smoke in the combustion products. When the smoke is bad enough, it will be completely opaque in the visible region. In the infrared region, the loss is much less.

When a flame is clean burning or air rich, the combustion products are transparent in both the infrared and visible regions. In the ultraviolet region, the combustion products of any flame, rich or lean, are opaque. These differences are important in applications involving more than one burner in a combustion chamber.

The following characteristics are those used for detecting a flame other than temperature:

1. Visible light radiation
2. Infrared light radiation
3. Ultraviolet light radiation
4. Ionized gas molecules in the flame

Visible Light Detectors

The Best Boiler Operator

Detection of flame by seeing the visible light, is, of course, the way man has detected flame since it was first observed. Since the radiation travels at the speed of light, the detection of flame by this method is not limited in speed.

There are two standard detectors used for sensing the flame in the visible region, the oldest of which is the photo-emissive photocell. This device is a glass-enclosed structure, which usually has been completely evacuated.

There are two elements within the envelope, a thin metal plate that is the cathode, and a collector assembly, which is the anode. When light strikes the cathode, electrons are emitted which are drawn to the anode. The current in this type of tube is very small and requires considerable amplification before it can operate a relay.

The other type of detector in the visible region is usually a smaller device that is constructed from an insulating plate covered with a deposit of cadmium sulfide that is mounted in a protective enclosure with a glass window. This device acts like a resistor, which is sensitive to light. When the cadmium sulfide detector is in the dark, the resistance of the element will be many megohms.

When it is exposed to a light from a flame, the resistance will drop to a few thousand ohms; and, in most devices, enough current will flow to energize a relay without further amplification. Flame detectors that operate in the visible region will also operate from other light in the visible region such as daylight or light from an incandescent of fluorescent lamp.

It is therefore necessary to make sure that they are used only in locations where they cannot see other sources of light. Also, they will respond to the visual light signal from hot refractory in the combustion chamber.

Exam Prep Course

Infrared Light Detectors

All flames radiate infrared energy, and since this radiation travels at the speed of light, the detection of flames by this method is not limited in speed. There is only one standard detector for use in the infrared region, and this is the lead sulfide detector.

This device is constructed by depositing a thin film of semiconductor material on an insulating surface. Two leads are attached to opposite ends of the film. The device operates like a resistor whose value of resistance changes with the amount of infrared energy falling on the surface of the detector.

These devices have dark resistance values from a few thousand ohms to a few megohms. When these detectors are exposed to light, they decrease in resistance depending on the intensity of the light. Their percentage change in resistance is not as great as in the cadmium sulfide detectors since the resistance of the element in the dark is much lower than the cadmium sulfide detector.

This requires a different type of amplifier for infrared detectors especially since the signal from a hot refractory is strong in the infrared region. An example of the infrared scanner is Type 48PT1 and 48PT2.

Ultraviolet Light Detectors

Ultraviolet radiation from a flame travels at the speed of light, therefore, the detection of flame by this method is not limited in speed.
The newer type of ultraviolet detector is constructed of two similar electrodes made from very clean wire of tungsten or molybdenum.

These operate from UV light falling on the wire electrodes,

which starts an avalanche of electrons. These tubes quench very fast. The fast quenching of these tubes allows them to be used on 50 or 60 cycle supply voltage, which provides automatic quenching.

Application of Photoelectric or Photocell Detectors in the Visible Region

Operation in the visible spectrum is limited to single burner applications because it is very difficult to differentiate between the burner being monitored and other burners which may be visible to the scanner since this would give a false indication of flame.

The normal application is for the smaller oil burners where the field of view is limited to the combustion flame itself. A blue gas flame will not operate any of the detectors that operate only in the visible region.

Detectors for use in the visible region should be located so that their field of view will include the flame at all times. It is important to consider the size of the flame under the worst condition, especially on those burners where the firing rate is variable. Usually, if the scanner can view the low fire flame, it will be satisfactory for the high fire. The best scanner position for the larger burner is on the end of a sight pipe mounted to the burner front. This should be located as near the burner as possible, and the line of sight should be as near parallel to the burner centerline as possible.

In some burners, it is possible to mount the scanner in the blast tube provided the swirl-inducing blades at the front of the blast tube do not obstruct the view.

Note: *The ambient temperature at the scanner should not exceed the rated temperature of the scanner.*

In most cases, this will not be a problem, but if the temperature of the scanner becomes too high, provisions should be made for connecting purge air to the scanner. This not only keeps the scanner cool, but also prevents the accumulation of dirt on the lens.

The sight pipe should be located so that the scanner is above the centerline of the flame so that the sight pipe will be angled downward to prevent dust and dirt from accumulating. The leads to the scanner should be arranged with a service loop so that the scanner can be removed for servicing. There should be clearance necessary for cleaning lenses or windows.

The field of view must include enough of the flame to provide a signal with a good margin of safety so that changes in the flame and accumulation of dust and dirt will not reduce the signal to the point where the control will shut down the burner. It is important that the field of view should not include hot refractory, which will be held in the control flame relay without a flame present.

Application of Infrared Detectors

Infrared detectors can be used with almost any single burner application. Some difficulty may be encountered with small gas pilots that have a limited infrared output. The infrared radiation passes readily through normal combustion products and dirty surfaces which makes them very dependable in a single burner application, but with multiple burners it is relatively easy for the scanner to see the adjacent burner.

Discrimination then becomes a problem of limiting the field of view of each scanner in the installation. Infrared detectors are ideally suited to large oil burner installations.
When a single scanner is to be used for detecting both pilot and main flame, the field of view of the scanner must be aimed at the intersection of the pilot and the main flame. This is to insure that a satisfactory signal from the scanner

The Best Boiler Operator

indicates that there is a flame at a point, which will be sure to ignite the main flame.

If two scanners are used to monitor the pilot and the main flame, then the scanner for the main flame must be aimed so that it will not detect the pilot. The scanner for the pilot should be aimed so that it views the intersection of the pilot and the main flame exactly as above in the case of a single scanner for pilot and main flame. The field of view of the scanners should be arranged as nearly as possible to avoid sighting hot refractory since this will reduce the sensitivity of the scanner.

The ambient temperature at the scanner should not be allowed to exceed 125 degrees F. If the temperature becomes a problem because of heat conducted up the sight pipe, it may be controlled by using insulating tubes or by providing purge air to the sight pipe. Providing purge air has the additional advantage of keeping dirt and dust out of the line of sight. If purge air is used or if the combustion chamber operates under higher than atmospheric pressure, it will be necessary to use a sealing union which has a pressure-tight window. The scanner sighting tube should be aimed in a downward direction so that dirt will not accumulate in the sight pipe and leads should be arranged with a service loop so that the scanner can be removed for cleaning or servicing.

Application of UV Detectors

All flames produce sufficient UV for ultraviolet detection. Even those flames completely invisible to the eye are easily seen with a UV detector. Since the combustion products of all flames are opaque in the UV region, this detector is well suited to multiple burner installations. A scanner will not detect the ultraviolet radiation from an adjacent flame because its combustion products will block the radiation. The ultraviolet detector is ideally suited to all gas burners or combination gas-oil burners as well as all multiple burners including powdered coal.

The area to scan with an ultraviolet detector is within the first 1/3 length of the flame since this is the major source of UV and also since, as pointed out earlier, the combustion products of the flame are opaque in the ultraviolet region. Do not try to scan at the outer fringes of the flame. The field of view should include the intersection of the pilot and the main flame so that a signal from the scanner will insure that the pilot is in the proper position to ignite the main flame.

Since the radiated energy from an electric spark igniter is very rich in the UV region, the field of view should be aimed so that it does not see an electric spark igniter nor any reflector that is close to the spark. Another very effective method of avoiding the signal from an electric ignition spark is to disconnect power from the ignition transformer before proving main flame. This not only eliminates the source of the interfering radiation but also insures a good stable pilot flame before turning on the main fuel valve.

The ambient temperature where the scanner is located must not be higher than 212 degrees F. Scanner cooling is possible with an insulated coupling or nipple or cooling air supplied to the sight pipe or by a sealing window.

The sealing window or lens assembly prevents leakage of hot gases when the combustion chamber is operated at higher pressure than atmospheric. The window is used when normal signal level is produced by the flame lens assembly when the UV level is too low for reliable operation.

Note: *It should be kept in mind that any windows in the UV scanner assembly must be made of special UV transmitting glass or quartz. Ordinary window glass or heat resisting glass will not transmit any UV radiation.*

The Best Boiler Operator

The sight pipe should be arranged to slant downward so that dirt and dust will not collect in the sight pipe. The leads to the scanner should be arranged with a service loop so that the scanner may be removed for cleaning or servicing.

The field of view should be large enough so that the signal produced by the scanner has a margin of safety to allow for changes in the flame and some accumulation of dust on the scanner window or lens.

Applications Requiring Self Checking
Ordinary burners used for heating and industrial processes normally cycle on and off many times during a day or week and this sequence are used to test the condition of the flame sensor and its amplifier.

A failure that causes the flame relay to energize before the fuel valve is opened will prevent the burner from being started. This important safety check does not occur if the burner operates continuously. In those applications where a burner operates 24 hours a day for periods of seven days or more, it is recommended that either the flame failure control be tested daily or some form of self-checking control should be used.

For ultraviolet sensors, this is accomplished by placing a shutter in front of the detector. When the shutter is closed, the sensor and amplifier must show a no-flame condition and when the shutter is open a flame-on condition must be detected.

This checking operation is repeated every six seconds. When reading the flame signal of a self-checking UV system, it will be noted that the signal level drops when the shutter is closed. The signal actually drops to zero, but a time delay in the meter circuit prevents the meter from dropping to zero.

For infrared sensors, the natural fluctuation of the flame

Exam Prep Course

radiation is used to actuate a special Autocheck control unit. The fluctuations are used to charge a capacitor network that operates a flame relay. Thus, a short circuit or open circuit in the scanner or aiming the scanner at a steady light source will not energize the flame relay. Also, any failure in the flame amplifier that prevents the fluctuating signal from charging the capacitor network will de-energize the flame relay.

Optical Principles

An understanding of basic optical principles will help solve three types of application problems.
1. Avoiding an unwanted signal such as:
    a. An adjacent flame from another burner
    b. Electric spark igniter (UV detector)
    c. Hot refractory (visible light detector)
    d. Light (visible light detector
2. Defining the exact area of a flame being monitored, such as the junction of the pilot and the main flame. This is necessary in order to insure that detection of the pilot will show that at least the minimum size required to light the main flame.
3. Increasing the amount of signal from a flame to insure reliable hold-in of the flame relay.
    a. Problems 1 and 2 are generally handled by controlling the field of view and problem 3 is handled either by increasing the field of view controlling the line of sight or by gathering more light with a lens arrangement. The different methods that can be used are described in this section.

This section covers the basic principles of optics as they apply to the use of flame detectors that operate from radiant energy. We use the term radiant energy because we are covering not only the visible light that we can see but also light in the infrared and ultraviolet regions. This radiant energy travels at the speed of light, which is 186,000 miles per second. The radiation also travels in straight lines and will pass through transparent materials.

Anything that is obviously transparent to the eye will be transparent to detectors, which operate in the visible region. The normal material for lenses and windows in the visible region is glass. Many transparent materials, such as ordinary glass and most plastics will not transmit in the ultraviolet regions. The only two common materials used in the ultraviolet region are fused quartz and a special UV transmitting glass. In the infrared region, some materials are transparent to the infrared but not in the visible region. The normal optical material used for flame detectors in the infrared region is heat-resisting glass.

The amount of light received by a detector varies with the intensity of the light source and with the distance from the light source. Since the signal from the detector in most cases depends on the amount of light received by the detector, it will be necessary to get more light on the detector if an installation does not provide sufficient margin of signal for trouble-free operation.

In the case of ultraviolet detection, as has been shown, an increased amount of light falling on the detector will not necessarily increase the signal. This is because most of the UV is generated in the first one-third of the flame and because of light absorption by the combustion products. In this case and in other cases involving discrimination or background interference, it will be necessary to control the field of view of the scanner.

The amount of light received by the detector varies with the distance from the source of light by a ratio, which depends on the optical arrangement. If the source of light is of a very small size, such as a candle or small gas pilot, and the detector is mounted at a relatively great distance, the light received by the detector will follow the "square law".

Exam Prep Course

The optical term for a small light source is a point source. The light from a point source radiates in all directions. Only the narrow bundle of light rays that falls on the sensitive area of the receiver will generate a signal.

Notice that a change in the position of the detector will affect the amount of light falling on the detector to a great extent; for instance, if d is doubled, the light received will be ¼ the intensity of the original light at the original distance.

The detector is frequently mounted in a sight pipe, but in this case this does not affect the signal received by the detector. Neither the length of the sight pipe nor its diameter affects the signal received. The "line of sight" is the direction in which the scanner is looking, or the centerline of its field of view.

The field of view of the scanner may be considered either the area, which the scanner can see, or the solid angle of unobstructed viewing of the detector. The end of the opening of the sight pipe limits the field of view so that the detector sees everything within this solid angle and nothing outside of the field of view. An aperture that limits the field of view is called the field stop.

If the detector were tilted so that the light falling on the sensitive surface arrives at an angle other than perpendicular, the amount of light received will decrease. If the angle of tilt is measured as the angle away from perpendicular, it will be found that the amount of light received is proportional to the cosine of the angle.

Therefore, when the angle becomes 90 degrees, the amount of light seen by the detector will be zero.
In systems without lenses, there are two methods for increasing the amount of light on a detector or the signal from it. These are: 1) Decrease the length of the sight pipe and 2)

The Best Boiler Operator

Increase the field of view.

Flame detectors operating from radiant energy can operate either intentionally or accidentally from reflected light, such as polished metal or a mirror. Rough surfaces are poor reflectors because they absorb some of the light and the rest is reflected in all different directions. The reflecting ability of materials varies in each of the spectral ranges. Infrared light reflects better from most materials than visible light, and the visible in turn reflects better than ultraviolet. In the case of ultraviolet, the reflection of an unwanted source such as an electrical spark is very poor, but since the spark is such a powerful light source in the ultraviolet region, even a small percentage of the reflected light can cause an unwanted signal.

Reflecting devices that are intended to help pick up a difficult signal should be considered under the worst operating condition. This would normally be with accumulation of dust and dirt and oily films, which will decrease the efficiency of reflection.

Lenses are frequently used with optical scanners to narrow the field of view or sometimes to increase the sensitivity of a scanner. A lens would, of course, have to be constructed from a material that would transmit the light intended.

How does a flame sensor work and how is it tested?

Ans. How the flame sensor works is by a small AC electrical charge going through the sensor and it uses the flame to carry the electrical power to ground. Burner surface to flame sensor is 4:1 ratio thus causing the AC current to change to DC current. If no flame present, dirty flame sensor, bad flame characteristics similar to flame lifting off or carbonizing there will be low or no micro amp draw and the control board will shut the flames off. After 3 to 5 tries the control board will lock the furnace out for 3 hours, which can be reset by

Exam Prep Course

cycling power or thermostat.

All furnaces have different micro amp readings some good at .8 while others need 6 you will need to find out the right micro amp reading for your furnace. The flame sensor has a coating on it to make it last a long time and will get worn off by over cleaning, so it is best to only clean rod when it is needed. If flame sensor requires cleaning more often than once a year it should be replaced, or check for other problems like vent stack temperature, gas pressure, or primary air need to be adjusted.

How to clean flame sensor:

1. First find the flame sensor.
   Most furnaces you will find a white wire located on the left side connecting the flame sensor to the circuit board.
   Watch the furnace ignites you will see spark or orange glow in front of one burner and the flame sensor should be on the opposite side, but I have seen them on the next burner from the igniter.
2. Shut power off to the furnace while you are working, you will find a light switch high on wall or in the ceiling.
3. Most furnaces have 1/4 inch screw that holds the sensor in place; any multi screwdriver without the bit will be 1/4 inch. Remove the screw(s) and disconnect wire
4. Once removed you can clean the sensor with a scotch-brite pad, if you do not have one then just use a clean rag for now. Do not use sandpaper or wire brush, as it will remove the protective coating and scratch the rod causing it to fail more often. Some people like to clean every year but I advise not to, It is best to clean only when needed since the protective coating will be removed with over cleaning
5. Once the rod is cleaned reconnect the wire and screws making sure the rod sits in the flame. Sometimes the burner and any grounding surface will get

The Best Boiler Operator
dirty and need to be cleaned as well.

Exam Prep Course
Chapter 5: Boiler Auxiliaries

Forced Draft Fans:

What is a Forced Draft Fan?

Ans. The Forced Draft Fan (FD Fan) is a fundamental part of most boiler systems and is the element responsible for creating draft inside of the boiler. The FD Fan provides the pressure and flow required to push air, fuel, and the resulting flue gases created from combustion through the boiler, catalysts, economizer, ductwork, and stack. FD Fans are manufactured in several different arrangements, depending on the size of the boiler. Auxiliary pieces that often accompany the FD Fan include Inlet Silencers, Flue Gas Mixing Boxes, Dampers, and Damper Actuators. FD Fans are designed to be driven by a Steam Turbine or an Electric Motor with an optional Variable Frequency Drive (VFD). Occasionally, the FD Fan is designed to use both of these prime movers in conjunction.

The FD fan is located at the front of the boiler to push air into the combustion chamber or furnace creating a positive pressure. If there are any leaks in the furnace area the combustion gases will leak out into the boiler room and surrounding area. This is what is meant by the furnace being pressurized.

The Best Boiler Operator

FDs are used in small to medium pressure boilers. The size of the FD must be large enough to force the combustion gases through the combustion space around the combustion dampers or baffles through the connecting ductwork and out the stack.

Induced Draft Fans:

What is an induced draft fan?

Ans. **Induced-draft**. An **induced-draft** burner uses a **blower** to pull air into the burner, and through the combustion chamber and heat exchanger. The **fan** then pushes the flue gases out through the vent. This creates a negative pressure in the furnace, and may create positive or negative pressure in the venting system.

What is the difference between a FD and an ID fan?

In a boiler arrangement, a forced draft fan will draw in air and force it into the combustion chamber of the boiler, where it mixes with the fuel being supplied. FD fans are typically used to regulate the proper amount of air-to-fuel ratio in an effort

Exam Prep Course

to maximize fuel efficiency and to minimize EPA-regulated emissions, such as NOx (Nitrogen Oxides).

Induced Draft fans are commonly used to draw flue gases from the combustion chamber and through the rest of the system to the stack. They help most to regulate the pressure inside of the boiler system.

In smaller boiler systems, the use of either an FD or ID fan is sufficient, but in large operations, both FD and ID fans are used together to regulate all of the above-mentioned factors simultaneously.

In some cases, additional fans are also added to the system, such as "Primary Air" fans and "Booster" fans. PA fans are typical in coal-burning operations, as more air is needed for the complete combustion of coal than an FD fan alone can provide. PA fans also provide transport and tempering air that is used to move pulverized coal from storage to the furnace. "Booster" fans are also typical of coal operations these days, as all of the additional equipment associated with "clean coal technology" adds a greater amount of pressure resistance in the exit path of flue gases, which a booster fan helps overcome.

What are some problems encountered with an ID fan?

Ans. Because an induced draft fan is located near the stack or in the flow of combustion gases as it sweeps through it the corrosive and abrasive material from the combustion gases can clog or erode the fan blades and housing. This is especially true in pulverized coal arrangements.

Boiler Feed Water Tank:

The Best Boiler Operator

What is a boiler feed tank? What is its purpose?

Ans. The boiler feed tank is usually an assembly of a holding tank for the returning condensate water coming back to the boiler room by condensate pumps. The feedwater controls are usually attached to the tank and when the boiler water level drops the feedwater regulator sends a signal to the controller. The valve opens and the feedwater pumps send water from the tank to the boiler. The water inside the tank is a mixture of condensate and make-up water added to the system to replace water lost through boiler blowdown and any leaks in the return lines.

The purpose is to hold treated condensate and allow for adding fresh treated make-up to the boiler when the boiler calls for more water.

What issues can be found with the boiler tank?

Ans. The tank can become corroded. If it becomes corroded from leaks or poor water treatment the tank will become corroded and eat away at the tank metal which will cause more leaks out and air into the system. Air into the system is very corrosive because of the moisture in the air acting on the

Exam Prep Course

metal causes rapid oxidation.

What is the gas train?

Ans. The gas train are the components needed to safely and efficiently deliver and burn natural gas to release the potential stored chemical energy in the fuel so that it can be used as heat energy to make steam in the boiler.

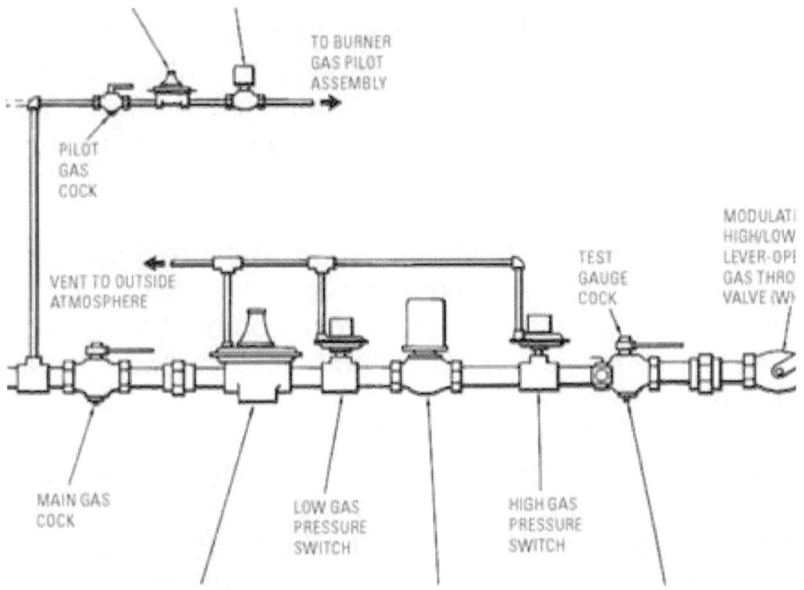

Refer to the diagram above. Every component on the gas train serves a purpose and is there to help the boiler furnace or combustion area receive fuel in the right amount under the right pressure at the right time. Starting from left to right you see the gas supply line. That is the line coming from your local gas supplier.

Drip leg: coming off the main line to drain or remove any moisture that may have been brought into the fuel supply line and would travel through the other components causing damage if not removed.

The Best Boiler Operator

Gas Pilot Line: the line coming off the t-connection for the gas pilot line. A small portion of gas is tapped off before the main gas cock and routed through the pilot gas cock and valves.

Main Gas Cock: a shut-off valve for the gas train at that point.

Gas Pressure Regulator: which controls or modulates the flow of gas so that the pressure coming through the system and at the tip is correct for the best combustion which in turn releases the most heat to be transferred to the boiler water through the heating surfaces.

Low Gas Pressure Switch: if the gas pressure drops too low for combustion the low gas pressure switch will shut off the flow of gas to that point preventing combustion from occurring. This valve is automatic and has to be manually reset if tripped.

Automatic gas safety: this valve opens after all of the other devices and components before it say that it is fine for it to open and allow gas to flow through. These devices are all in communication and operate relative to each other. They are sequential. One has to prove that its conditions are met before sending gas onward for combustion at the burner.

High gas pressure switch: If the gas pressure is too high coming through the train the high pressure gas switch shuts down the flow of gas.

Modulating firing lever: opens and closes the flow of gas to the burner based on the boiler load.

What is an Oil Train?

Ans. The gas train are the components needed to safely and

efficiently deliver and burn fuel oil to release the potential stored chemical energy in the fuel so that it can be used as heat energy to make steam in the boiler.

The Best Boiler Operator
Chapter 6: Boiler Accessories

What is a water column?

Ans. A cast iron or steel device depending on the boiler operating pressure that is directly connected to the boiler to both the steam and water side. Any valves located between the water column and the boiler must be permanently open and of the OS&Y type.

The water column serves as the middleman between the boiler and the gauge glass. The gauge glass cannot be connected directly to the boiler. Because of the violent boiling of the water that occurs inside a boiler during operation a gauge glass would not give the operator a true indication of the current water level if directly connected. Therefore the water column is necessary. It communicates with the boiler in the sense that the boiler lets the column know where the water level is but does not talk to the gauge glass. The gauge glass has to get the water level information from the water column.

Exam Prep Course

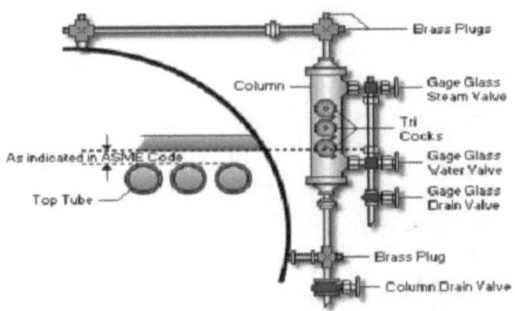

Water columns have a blowdown line to remove sediment that may get trapped inside. To keep the lines clear it is a great practice to blowdown the water column once a shift. **What are tri-cocks?**

Ans. Tri-cocks are connected directly on the water column and are three valves that when opened gives the operator the true water level of the boiler. Depending on the height of the boiler the tri-cocks may be opened by a chain from the floor. The top tri-cock when opened releases steam. It does so because it is connected to the top of the steam space of the boiler drum. There should not be anything in this space but wet steam. The middle cock when opened should release a steam/water mixture. It does so because it is located at the Normal Operating Water Level or NOWL. If you get steam out of the middle cock that will tell an operator that there is a low water condition inside the boiler that needs immediate attention. The bottom cock releases water because it is located under the NOWL. If you have steam coming out of the bottom cock then you have a dangerously low water condition and you should shut the boiler down immediately to allow to cool.

THE BEST BOILER OPERATOR

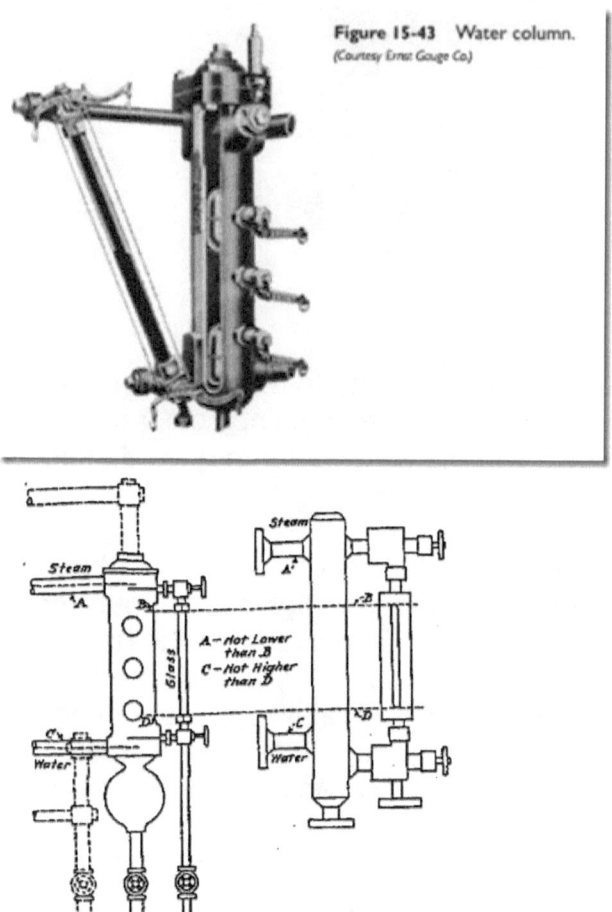

Figure 15-43 Water column.
(Courtesy Ernst Gauge Co.)

The tri-cocks are a secondary but definite means of determining the water level inside the boiler.

What is a steam gauge?

Ans. A device that tells the operator of the pressure operating inside the boiler. The gauge is graduated to 1 ½ times the boiler's MAWP. Meaning if the boiler MAWP is 30 pounds per square inch or ps then the gauge will have a maximum reading pressure on its face of 45 psi.

Exam Prep Course

The gauge is directly connected to the highest part of the steam side of a boiler drum. There is a siphon loop that looks like a pig's tail located between the inlet of the gauge and the steam connection from the boiler to the gauge. This siphon loop contains water that protects the gauge internal parts from the heat of the steam when the boiler is operating. There can be no valves between the gauge and the boiler.

How does a boiler feed pump work?

All boiler feed pumps operate by creating a pressure differential that allows a fluid to flow since fluids flow from a high pressure to a lower pressure. A pump creates that pressure difference.

What are some types of boiler feed pumps?

Positive Displacement/piston:

A duplex pump is a pump typically found in older boiler, wastewater treatment plants and oil fields. It consists of two pumps that alternate the pumping process, thereby allowing it to be more efficient than a single version. This pump has a higher flow rate due to its not having a dead spot in the pump stroke. As one pump is completing its stroke cycle, the other is beginning its stroke — maintaining maximum pumping action without a break in the cycle.

THE BEST BOILER OPERATOR

Non-positive displacement/Centrifugal:

A centrifugal pump is one of the simplest pieces of equipment in any process plant. Its purpose is to convert the energy of a prime mover (an electric motor or turbine) first into velocity or kinetic energy and then into pressure energy of a fluid that is being pumped. The energy changes occur by virtue of two main parts of the pump, the impeller and the volute or diffuser. The impeller is the rotating part that converts driver energy into the kinetic energy. The volute or diffuser is the stationary part that converts the kinetic energy into pressure energy. **Note**: All of the forms of energy involved in a liquid flow system are expressed in terms of feet of liquid i.e. head.

The process liquid enters the suction nozzle and then into eye (center) of a revolving device known as an impeller. When the impeller rotates, it spins the liquid sitting in the cavities between the vanes outward and provides centrifugal acceleration. As liquid leaves the eye of the impeller a low-pressure area is created causing more liquid to flow toward the inlet. Because the impeller blades are curved, the fluid is pushed in a tangential and radial direction by the centrifugal force. This force acting inside the pump is the same one that keeps water inside a bucket that is rotating at the end of a string.

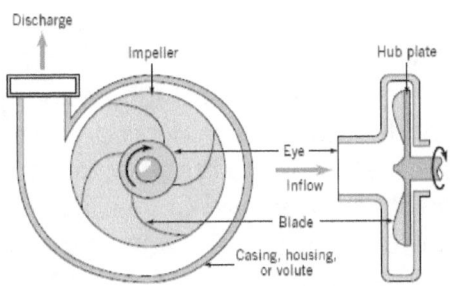

The key idea is that the energy created by the centrifugal force is kinetic energy. The amount of energy given to the liquid is proportional to the velocity at the edge or vane tip of the impeller. The faster the impeller revolves or the bigger the impeller is, then the higher will be the velocity of the liquid at the vane tip and the greater the energy imparted to the liquid. This kinetic energy of a liquid coming out of an impeller is harnessed by creating a resistance to the flow. The first resistance is created by the pump volute (casing) that catches the liquid and slows it down. In the discharge nozzle, the liquid further decelerates and its velocity is converted to pressure.

General Components of Centrifugal Pumps

A centrifugal pump has two main components:

1. A rotating component comprised of an impeller and a

shaft
2. A stationary component comprised of a casing, casing cover, and bearings.

Stationary Components.

Casing Casings are generally of two types: volute and circular. The impellers are fitted inside the casing.

1. Volute casings build a higher head; circular casings are used for low head and high capacity. o A volute is a curved funnel increasing in area to the discharge port.. As the area of the cross-section increases, the volute reduces the speed of the liquid and increases the pressure of the liquid. One of the main purposes of a volute casing is to help balance the hydraulic pressure on the shaft of the pump. However, this occurs best at the manufacturer's recommended capacity. Running volute-style pumps at a lower capacity than the manufacturer recommends can put lateral stress on the shaft of the pump, increasing wear-and-tear on the seals and bearings, and on the shaft itself. Double-volute casings are used when the radial thrusts become significant at reduced capacities.

2. Circular casing have stationary diffusion vanes surrounding the impeller periphery that convert velocity energy to pressure energy. Conventionally, the diffusers are applied to multi-stage pumps. The casings can be designed either as solid casings or split casings. Solid casing implies a design in which the entire casing including the discharge nozzle is all contained in one casting or fabricated piece. A split casing implies two or more parts are fastened together. When the casing parts are divided by horizontal plane, the casing is described as horizontally split or axially split casing. When the split is in a vertical plane perpendicular to the rotation axis, the casing is described as

Exam Prep Course

vertically split or radially split casing. Casing Wear rings act as the seal between the casing and the impeller.

Suction and Discharge Nozzle.

The suction and discharge nozzles are part of the casings itself. They commonly have the following configurations.

1. End suction/Top discharge - The suction nozzle is located at the end of, and concentric to, the shaft while the discharge nozzle is located at the top of the case perpendicular to the shaft. This pump is always of an overhung type and typically has lower NPSHr because the liquid feeds directly into the impeller eye.

2. op suction Top discharge nozzle -The suction and discharge nozzles are located at the top of the case perpendicular to the shaft. This pump can either be an overhung type or between-bearing type but is always a radially split case pump.

3. Side suction / Side discharge nozzles - The suction and discharge nozzles are located at the sides of the case perpendicular to the shaft. This pump can have either an axially or radially split case type.

Seal Chamber and Stuffing Box.

Seal chamber and Stuffing box both refer to a chamber, either integral with or separate from the pump case housing that forms the region between the shaft and casing where sealing media are installed. When the sealing is achieved by means of a mechanical seal, the chamber is commonly referred to as a Seal Chamber. When the sealing is achieved by means of packing, the chamber is referred to as a Stuffing Box. Both the seal chamber and the stuffing box have the primary function of protecting the pump against leakage at the point where the shaft passes out through the pump pressure casing. When the pressure at the bottom of the chamber is below atmospheric, it prevents air leakage into the pump. When the pressure is above atmospheric, the chambers prevent liquid leakage out of the pump. The seal chambers and stuffing boxes are also provided with cooling or heating arrangement for proper temperature control. Figure B.06 below depicts an externally mounted seal chamber and its parts.

Rotating Components.

1. Impeller The impeller is the main rotating part that provides the centrifugal acceleration to the fluid. They are often classified in many ways. o Based on major direction of flow in reference to the axis of rotation ? Radial flow ? Axial flow ? Mixed flow o Based on suction type ? Single-suction: Liquid inlet on one side. ? Double-suction: Liquid inlet to the impeller symmetrically from both sides.

o Based on mechanical construction Closed: Shrouds or sidewall enclosing the vanes. ? Open: No shrouds or wall to enclose the vanes. ? Semi-open or vortex type.

Closed impellers require wear rings and these wear rings present another maintenance problem. Open and semi-open impellers are less likely to clog, but need manual adjustment to the volute or back-plate to get the proper impeller setting and prevent internal re-circulation. Vortex pump impellers are great for solids and "stringy" materials but they are up to 50% less efficient than conventional designs. The number of impellers determines the number of stages of the pump. A single stage pump has one impeller only and is best for low head service. A two-stage pump has two impellers in series for medium head service. A multi-stage pump has three or more impellers in series for high head service. o Wear rings: Wear ring provides an easily and economically renewable leakage joint between the impeller and the casing. clearance becomes too large the pump efficiency will be lowered causing heat and vibration problems. Most manufacturers require that you disassemble the pump to check the wear ring clearance and replace the rings when this clearance doubles.

2. Shaft The basic purpose of a centrifugal pump shaft is to transmit the torques encountered when starting and during operation while supporting the impeller and other rotating parts. It must do this job with a deflection less than the minimum clearance between the rotating and stationary parts.

Shaft Sleeve (Figure B.08): Pump shafts are usually protected from erosion, corrosion , and wear at the seal chambers, leakage joints, internal bearings, and in the waterways by renewable sleeves. Unless otherwise specified, a shaft sleeve of wear, corrosion, and erosion resistant material shall be provided to protect the shaft. The sleeve shall be sealed at one end. The shaft sleeve assembly shall extend beyond the outer face of the seal gland plate. (Leakage between the shaft and the sleeve should not be confused with leakage through the mechanical seal).

Coupling: Couplings can compensate for axial growth of the shaft and transmit torque to the impeller. Shaft couplings can be broadly classified into two groups: rigid and flexible. Rigid couplings are used in applications where there is absolutely no possibility or room for any misalignment. Flexible shaft couplings are more prone to selection, installation and maintenance errors. Flexible shaft couplings can be divided into two basic groups: elastomeric and non-elastomeric ? Elastomeric couplings use either rubber or polymer elements to achieve flexibility. These elements can either be in shear or in compression. Tire and rubber sleeve designs are elastomer in shear couplings; jaw and pin and bushing designs are elastomer in compression couplings. ? Non-elastomeric couplings use metallic elements to obtain flexibility. These can be one of two types: lubricated or unlubricated. Lubricated designs accommodate misalignment by the sliding action of their components, hence the need for lubrication. The non-lubricated designs accommodate misalignment through flexing. Gear, grid and chain couplings are examples of non-elastomeric, lubricated couplings. Disc and diaphragm couplings are non-elastomeric and non lubricated.

Auxiliary Component.

Auxiliary components generally include the following piping systems for the following services:
- Seal flushing
- Cooling
- quenching systems
- Seal drains and vents
- bearing lubrication
- cooling systems
- Seal chamber or stuffing box cooling
- heating systems
- Pump pedestal cooling systems

Auxiliary piping systems include:
- tubing, piping, isolating valves, control valves, relief valves, temperature gauges and thermocouples, pressure gauges, sight flow indicators, orifices, seal flush coolers, dual seal barrier/buffer fluid reservoirs, and all related vents and drains.

All auxiliary components shall comply with the requirements as per standard codes like API 610 (refinery services), API 682 (shaft sealing systems) etc.

Definition of Important Terms: The key performance parameters of centrifugal pumps are capacity, head, BHP (Brake horsepower), BEP (Best efficiency point) and specific speed. The pump curves provide the operating window within which these parameters can be varied for satisfactory pump operation. The following parameters or terms are discussed in detail in this section.

Exam Prep Course

Capacity/ Head
- Significance of using Head instead of Pressure
- Pressure to Head Conversion formula
- Static Suction Head, hS o Static Discharge Head, hd
- Friction Head, hf

Vapor pressure Head, hvp
- Pressure Head, hp
- Velocity Head, hv
- Total Suction Head HS o Total Discharge Head Hd
- Total Differential Head HT

NPSH
- Net Positive Suction Head Required NPSHr
- Net Positive Suction Head Available NPSHa
- Power (Brake HorsePower, B.H.P) and Efficiency (Best Efficiency Point, B.E.P)   Specific Speed

Capacity

Capacity means the flow rate with which liquid is moved or pushed by the pump to the desired point in the process. It is commonly measured in either gallons per minute (gpm) or cubic meters per hour (m3/hr). The capacity usually changes with the changes in operation of the process. For example, a boiler feed pump is an application that needs a constant pressure with varying capacities to meet a changing steam demand.

The capacity depends on a number of factors like: Process liquid characteristics i.e. density, viscosity Size of the pump and its inlet and outlet sections Impeller size Impeller rotational speed RPM Size and shape of cavities between the vanes Pump suction and discharge temperature and pressure conditions For a pump with a particular impeller running at a certain speed in a liquid, the only items on the list above that can change the amount flowing through the pump are the pressures at the pump inlet and outlet.

The effect on the flow through a pump by changing the outlet pressures is graphed on a pump curve. As liquids are essentially incompressible, the capacity is directly related to the velocity of flow in the suction pipe.

Head

Significance of using the "head" term instead of the "pressure" term

The pressure at any point in a liquid can be thought of as being caused by a vertical column of the liquid due to its weight. The height of this column is called the static head and is expressed in terms of feet of liquid. The same head term is used to measure the kinetic energy created by the pump.

In other words, head is a measurement of the height of a liquid column that the pump could create from the kinetic energy imparted to the liquid. Imagine a pipe shooting a jet of water straight up into the air, the height the water goes up would be the head. The head is not equivalent to pressure.

Head is a term that has units of a length or feet and pressure has units of force per unit area or pound per square inch. The main reason for using head instead of pressure to measure a centrifugal pump's energy is that the pressure from a pump will change if the specific gravity (weight) of the liquid changes, but the head will not change.

Since any given centrifugal pump can move a lot of different fluids, with different specific gravity, it is simpler to discuss the pump's head and forget about the pressure. So a centrifugal pump's performance on any Newtonian fluid, whether it's heavy (sulfuric acid) or light (gasoline) is described by using the term 'head'. The pump performance curves are mostly described in terms of head.

Exam Prep Course

What is a fusible plug?

It is a very important warning device of a steam boiler, which protects the fire tube boiler against overheating. It is fitted on the fire box brown plate or over the combustion chamber. The fusible plug consists of two hollow guns and one conical plug shown as figure. A hollow gun metal body is screwed to the fire box crown plate of boiler. Another hollow gun metal is screwed to the first body. Third plug is made from copper is locked with the second plug by pouring metal into the grooves provided on the both plugs.

In normal working condition, the upper surface of fusible plug is covered with water which keeps the temperature of the plug below its melting point while the other end of the plug is exposed to fire or hot gases. The low melting point (tin or lead) does not melt until the upper surface of plug is submerged in water. But in the case of water level in the boiler falls below the danger levels, the fusible plug uncovered by the water and get exposed to steam. This overheats the plug and the fusible metal having low melting point which melts quickly. Thus the third plug drops down and second hollow gun became open, the steam rushes into the furnace and puts out the fire.

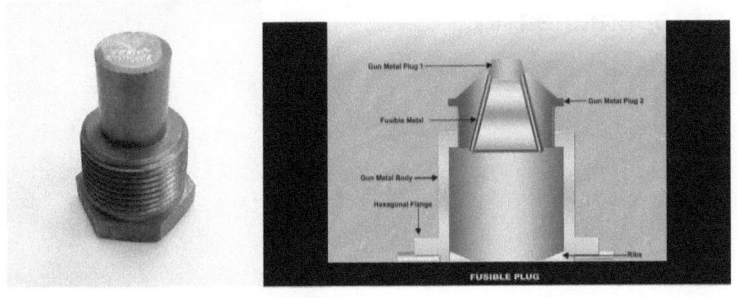

The Best Boiler Operator

What is a steam separator?

Steam separator has main function to separate water and steam and this equipment is usually located in steam drum. Water surface in steam drum is turbulent, so make it easy to mix between steam and water. The principle of steam separator is make steam flow is changed in any direction. Because of the density of steam is lighter than water make steam can be distributed easier than water. The water droplet which has higher density will be separated and dropped from steam. Moisture will be removed by steam separator to eliminate damage and erosion if water or wet steam is distributed to the steam line.

There are some types of steam separator. For small scale steam boiler, steam separator consists of dry pipe which has a lot of holes at the top and two holes at the bottom half. The mixture steam-water is directed through the top half holes dry pipe, turbulent moving force the mixture to separate between water and steam. Steam will flow to steam line and water will drop through bottom holes.

For big scale boiler which has complexity equipment, steam separator method use centrifugal force for better result. The mixture steam-water is forced to move around the cyclone and make the rotation. The more turbulent moving force the mixture separate easily.

Exam Prep Course

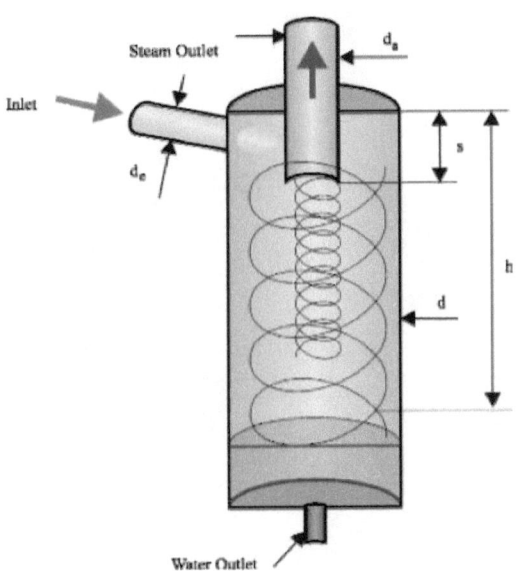

What is a steam baffle?

In the water tube boiler the upper drum (**Boiler steam drum**) provides for separation of steam from water. It also provides liquid holdup capacity (typically 10 to 60 seconds) to allow for a dynamic response to load changes without losing liquid in the downcomer and riser tubes. The size of the steam drum is determined by the volume required for a clean separation between steam and water to produce a dry saturated steam, it also provides an adequate amount of steam storage. The steam drum is also called the upper drum.

The Best Boiler Operator

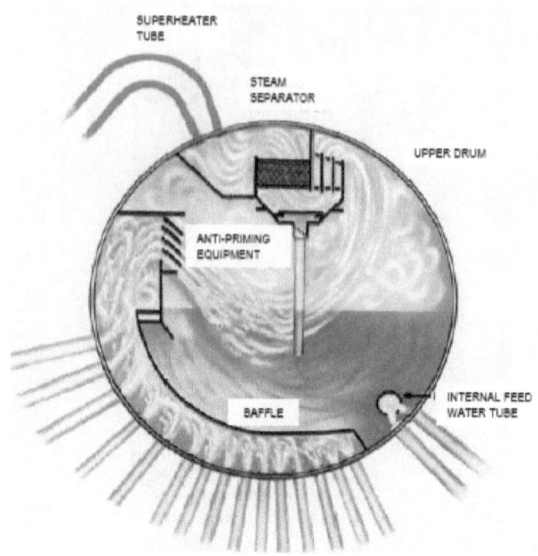

Steam drum internals for typical type of boiler are shown. The internals include separation devices to assist in separating small droplets of boiler water from the steam. Boiler water in the steam will cause scaling of the superheater and steam turbines. Baffles, chevron types separators and demister pads are used to minimize boiler water carryover. The demisters coalesce the smaller drops into larger drops so that they will settle back into the water phase. An alternate separation uses a cyclone separators as the first separating device followed by demister pads. Saturated steam from the steam drum contain no more than 0.5% moisture and 0.5 ppm solids.

Other steam drum internals include feedwater piping, blowdown piping, and chemical injection piping. The feedwater inlet must be baffled so that changes in BFW Boiler Feed Water rate do not set up waves that affect level measurement.

The steam drum plate into which the downcomer and riser

tubes are connected is thicker than the rest of the drum. The tubes are rolled into this plate like the tubes in an exchanger are rolled into the tube sheet. Expanded tube connections are generally suitable for pressures up to 1200 psia. A typical drum plate was almost 3.5" thick for a design pressure at 800 psig. Sometimes the tubes are also back welded, especially in high pressure boilers. Tubes may be back welded to repair leaks.

What is a manhole? Handhole?

Manholes are the access door provided in the boiler outer casing or in the shell of a closed pressure vessel for human and material entry into the otherwise closed (confined) space.

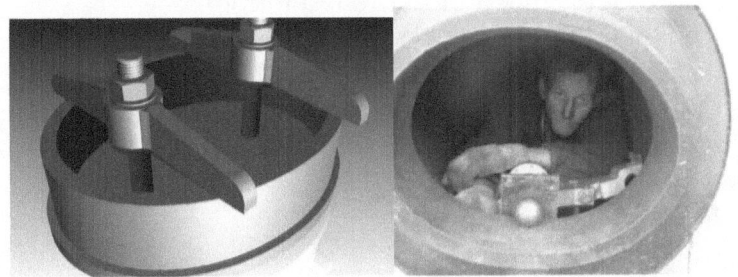

Manholes providing on the shell of the high pressure vessel like boiler drum or high pressure heaters are either circular or elliptical.

Drum manhole plates are normally placed inside the drum. Internal pressure from the vessel (inside) and bolt tightening pressure keeps the manhole plate in place. Unequal dimensions along the axes the plate and hole facilitates the plate to enter inside the vessel.

The Best Boiler Operator

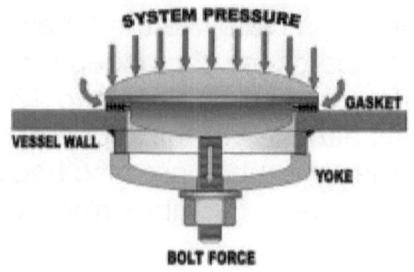

Manhole sizes

Normal size of a manhole is around 450 mm square or circular. In the case of elliptical minor axis will be 400 mm.
Precautions:
Manhole provides access to a confined space. Confined space entry requires certain precautions.
1. a. Before entry ensure that enough oxygen is available and no toxic gases present.
2. b. Enough lighting should be made available.
3. C. Landing area should be clear and should not be more deep more than the leg length.
4. d. Holding rods should be made available on both inside and outside preferably.
5. Manholes are meant as the access as well as material entry gate in a confined space.

What is a blowdown line?

A steam boiler evaporates liquid water to form steam, or gaseous water, and requires frequent replenishment of boiler feedwater for the continuous production of steam required by most boiler applications. Water is a capable solvent, and will dissolve small amounts of solids from piping and containers including the boiler. Continuing evaporation of steam concentrates dissolved impurities until they reach levels potentially damaging to steam production within the boiler. Without blowdown, impurities would reach saturation levels and begin to precipitate within the boiler. Impurity concentrations are highest where steam is being produced near the heat exchange surfaces. Precipitation would be expected to occur in the form of scale deposits on the heat exchange surfaces. Scale deposits thermally insulate heat exchange surfaces initially decreasing the rate of steam generation, and potentially causing boiler metals to reach failure temperatures.

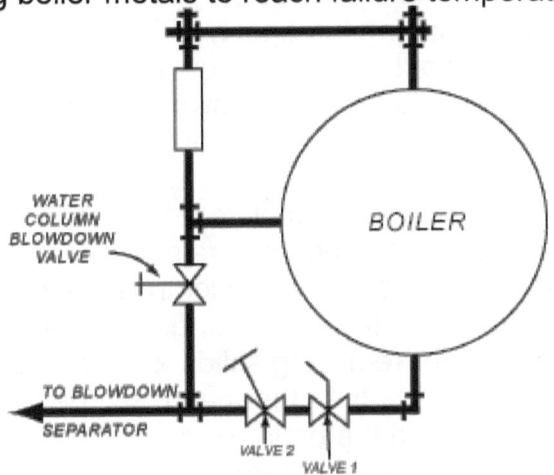

What is the sequence in blowing down a boiler?

Typically boiler bottom drains have a knife valve and a gate valve. The valve(s) should be piped to a discharge point that is safe. The boiler blowdown valve sequence involves ensuring both valves are closed, then opening the knife valve, then opening and closing the gate valve. Otherwise, the general advice is to open the blowdown valve closest to the boiler first, and close it last, with the actual drainage/blowing down happening from the valve furthest from the boiler. Repeating the process several times moves sludge toward the drain line.

You may also use a boiler blowdown valve sequence of opening the quick opening valve, or the one that controls the release of water, until the water in the boiler is reduced by half. Then, close all of the valves that are open and the water will blow down.

For a skimmer or surface blowdown, a needle valve is typically used, or a flow throttling valve. Boiler water goes through this valve into a skimmer pipe. The specific method and schedule for boiler blowdown you choose depends on many factors, including the quality of the make-up water. Generally, a bottom blowdown should be completed at least once a day, and a surface blowdown can actually be automated.

There are a few ways to complete a boiler blowdown. Water can be drained from the bottom drain valve, or a skimmer drain valve. Dissolved solids concentrate the most about six to eight inches below the water surface, making the skimmer drain valve ideal for removing solids without reducing as much boiler water. On the other hand, sludge in the boiler is at its heaviest toward the bottom, so a bottom valve blowdown is still important for reducing buildup.

Why is blowing down a boiler so important?

Some boiler water treatments cause precipitation of impurities as insoluble particles anticipating those particles will settle to the bottom of the boiler before they become entrained in water circulating past the heat exchange surfaces. These water treatments often include compounds forming a sludge to entrap such particles; and boilers intended for such water treatment include a structure called a mud drum at the lowest part of the boiler. Bottom blowdown involves periodically opening valves in the mud drum to allow boiler pressure to force accumulated sludge out of the boiler. Similar blowdown connections at the bottom of water wall headers are blown down less frequently.

Several short blowdown events remove sludge more effectively than a single continuous blowdown. Shorter blowdown events cause less significant changes in boiler water level, and are safer during periods of high steam demand.

Bottom blowdown piping drains the lowest parts of the boiler so it can be used to drain the boiler for servicing. Bottom blowdown piping must be of a diameter large enough to minimize the risk of being plugged with baked sludge.

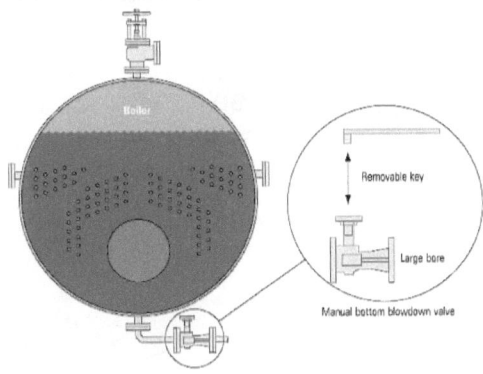

Modern boilers discharge bottom blowdown to a blowoff tank where the blowdown can flash and vent steam upwards without entraining water which might cause burns. A pipe near the bottom of the blowoff tank maintains a water level below the blowdown entry point and allows cooler water remaining from earlier blowdown events to drain from the tank first. Two bottom blowdown valves are often used in series to minimize erosion. One valve serves as the sealing valve, and the other as the blowdown valve. The sealing valve is customarily opened first and closed last. Both are opened rapidly and fully to minimize erosion on the seat and disk faces. Care is taken to avoid trapping scale or rust particles within the valve by reopening a valve to flush the particles through if resistance is encountered when attempting to close it. Bottom blowdown valves are often rebuilt or replaced whenever the boiler is taken out of service for maintenance

What is a steam injector?

A **steam injector** is typically used to deliver cold water to a

The Best Boiler Operator

boiler against its own pressure using its own live or exhaust steam, replacing any mechanical pump. This was the purpose for which it was originally invented in 1858 by Henri Giffard. Its operation was from the start intriguing since it seemed paradoxical, almost like perpetual motion, but its operation was later explained using thermodynamics. Other types of injector may use other pressurised motive fluids such as air.

Depending on the application, an injector can also take the form of an eductor-jet pump, a water eductor or an aspirator. An ejector operates on similar principles to create a vacuum feed connection for braking systems etc.

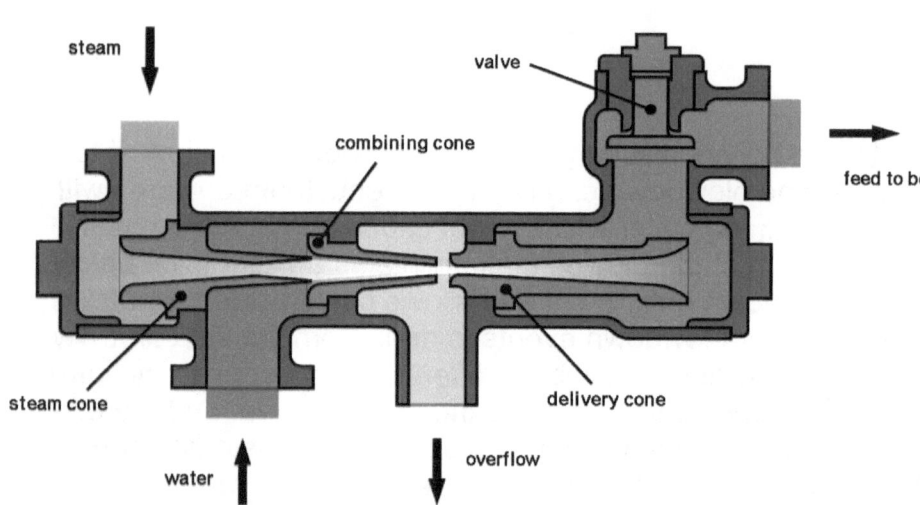

The injector consists of a body filled with a secondary fluid, into which a motive fluid is injected. The motive fluid induces the secondary fluid to move. Injectors exist in many variations, and can have several stages, each repeating the same basic operating principle, to increase their overall effect.

It uses the Venturi effect of a converging-diverging nozzle on a steam jet to convert the pressure energy of the steam to velocity energy, reducing its pressure to below atmospheric which enables it to entrain a fluid (eg. water). After passing through the convergent "combining cone", the mixed fluid is fully condensed releasing the latent heat of evaporation of the steam which imparts extra velocity to the water. The condensate mixture then enters a divergent "delivery cone" which slows the jet, converting kinetic energy back into static pressure energy above the pressure of the boiler enabling its feed through a non-return valve. Most of the heat energy in the condensed steam is returned to the boiler, increasing the thermal efficiency of the process. Injectors are therefore typically over 98% energy-efficient overall; they are also simple compared to the many moving parts in a feed pump.

Steam injector of a locomotive boiler The motive fluid may be a liquid, steam or any other gas. The entrained suction fluid may be a gas, a liquid, slurry, or a dust-laden gas stream.

The Best Boiler Operator
Chapter 7: Boiler Start-Up

A review of boiler startup procedures can be summarized in the following list. Your actual system may vary but this guide provides an outline of common items to check. For a dry start, the boiler will have to be filled with water, for a wet layup boiler, the water level may have to be dropped and chemistry adjusted before bringing online.

Typical Startup Process

1. Check all valves and place in their startup position
2. Open the sight gauge and water column high- and low-water shut-off valves. Make sure the water level safety controls are blown out.
3. Close the bottom blowdown valves, then open the upper drum vent valves.
4. Start filling with soft water.
5. Manually inject boiler water treatment chemicals including oxygen scavenger chemicals, so that the chemicals are added with the fill water.
6. Once full to the operating level, open the fuel system and fire the boiler. Carefully bring the pressure up to 10-15 PSIG, with the vent valve open. The boiler's warm-up curve should be strictly followed. The standard warm-up curve for a typical boiler is not to increase the boiler water temperature over 100°F per hour. Check your manufacturers' guidelines.
7. After the pressure reaches 10-15 PSIG, close the drum vent and slowly bring the boiler up to operating pressure.
8. Collect a boiler water sample and test for the proper chemical concentrations. Adjust as needed.

While this procedure is mainly focused on the facility operator, as a water treatment provider, you can be asked to comment on these procedures and also assist with the proper startup chemistry guidelines. Being familiar with the startup processes can help you in the overall servicing of your customers and being able to answer their questions.

In summary, the proper startup of a boiler system is one step in the many required throughout the year to keep the boiler operating efficiently. Be it the startup, shutdown and layup, or operating throughout the year, each of these work together to keep a system problem free for the long haul. Proper chemistry is very important, but so are the mechanical and operating procedures in the facilities that operate these boilers.

The Best Boiler Operator

Exam Prep Course
Chapter 8: Boiler Operation

Each year, hundreds of accidents are reported nationally involving steam and hot water heating boilers. Most are attributed to malfunctioning low water cutoffs, operator error, poor maintenance and/or corrosion. Properly functioning control or safety devices are absolutely essential. The only way you can be confident they will work when needed is to regularly perform required maintenance and testing.

What are these control and safety devices? Some of the more obvious ones are listed below. The full-text article has basic recommendations for testing and maintenance. Consult your boiler manufacturer, contractor, insurer, or state boiler authorities for any questions about detailed procedures and requirements.

Safety Valves
Often considered the primary safety feature on a boiler, the safety valve should really be thought of as the **last line of defense**. If something goes wrong, the safety valve is designed to relieve all the pressure that can be generated within the boiler. Keep in mind that the same conditions that make other safety devices malfunction can also affect the safety valve. Don't let testing and maintenance schedules slide.

Water Level Control and Low Water Fuel Cutoffs
These devices perform two separate functions, but are often combined into a single unit. This method is economical, providing both a water level control function and the safety feature of a low water fuel cutoff device. We recommend, however, that both steam and hot water boilers always have two separate devices — a primary and a secondary low water fuel cutoff. Many jurisdictions require two such devices on steam boilers.

The Fuel System
Failure to maintain the equipment in good working order could result in higher fuel costs, the loss of heat transfer or even a furnace explosion. Modern fuel systems are very complex assemblies, consisting of both electronic and mechanical components. Over a period of time many things may go wrong. Many users wisely con-

The Best Boiler Operator

tract with their gas company or oil service company to periodically check and maintain their burner equipment.

The Water Gage Glass

The importance of proper cleaning and maintenance of the water gage glass, or sight glass, cannot be stressed enough. The gage glass on a steam boiler enables the operator to visually observe and verify the actual water level in the boiler. But water stains and clogged connections can result in false readings. The glass may break or leak. Take the time to replace the glass, even if the boiler must be shut down. That inconvenience is nothing compared to the damage that may result from operating a boiler without a gage glass.

The Stack Temperature Gage

A stack temperature gage is normally installed on a boiler to indicate the temperature of the flue gas leaving the boiler. A high stack temperature indicates that the tubes may be getting a buildup of soot or scale. Also, the baffling inside the boiler may have deteriorated or burned through, allowing gases to bypass heat transfer surfaces in the boiler.

Boiler Logs Are Important

The majority of boiler accidents can be prevented. One of the most effective tools is the proper use of operating and maintenance logs. Boiler logs are the best method to assure a boiler is receiving the required attention and provide a continuous record of the boiler's operation, maintenance and testing. Because a boiler's operating conditions change slowly over time, a log is the best way to detect significant changes that may otherwise go unnoticed.

Most boiler problems don't occur suddenly. They develop slowly over a long period of time. So slowly, in fact, that you grow accustomed to the change without realizing it has taken place. Corrosion slowly builds up in the safety valve. Sediment collects in the float chamber on the connection lines of the low water fuel cutoff device. Scale accumulates on the waterside of your boiler tubes.

Exam Prep Course

Each year, hundreds of accidents are reported nationally involving steam and hot water heating boilers in businesses, public buildings and other facilities. The majority of these incidents are attributed to malfunctioning low water cutoffs, operator error, poor maintenance and/or corrosion. Properly functioning control or safety devices are absolutely essential for any boiler. The only way you can be confident they will work when called upon to do so is to regularly perform required maintenance and testing.

What are these control and safety devices? Some of the more obvious ones are discussed below, with basic recommendations for testing and maintenance. These are not the only items on a boiler that contribute to its proper operation, but they are some of the primary ones. This discussion of testing and maintenance procedures is not exhaustive — consult your boiler manufacturer, contractor, insurer, or state boiler authorities for any questions about detailed procedures and requirements.

Safety Valves

Often considered the primary safety feature on a boiler, the safety valve should really be thought of as the last line of defense. If something goes wrong, the safety valve is designed to relieve all the pressure that can be generated within the boiler. Although it is essential, a safety valve can also give you a false sense of security that encourages letting the testing and maintenance schedules slide. Keep in mind that the same conditions that make other safety devices malfunction can also affect the safety valve.

Every steam and hot water heating boiler must have at least one safety or safety relief valve of sufficient relieving capacity to meet or exceed the maximum burner output. The ability of a safety valve to perform its intended function can be affected by several things, such as internal corrosion or restricted flow, which can prevent the valve from functioning as designed. Internal corrosion is probably the most common cause of "freezing" or binding of safety/relief valves. This condition is generally caused by slight leaking or "simmering" due to improper seating of the valve disk and is a condition that should be corrected without delay.

To assure that a valve's mechanism will operate properly, the try-lever should be lifted once a month and the valve set pressure tested annually. If a valve will not operate or does not reseat properly when tested, the boiler must be shut down immediately and the valve repaired or replaced. The safety or safety relief valve must be set to open at or below the maximum allowable working pressure established by the manufacturer. This is the maximum pressure at which the designers have determined the boiler can be safely operated. The maximum allowable working pressure is listed on the required boiler nameplate or stamping.

It is not good practice to operate a boiler too close to the valve setting. Operating too close to the set pressure will cause these valves to leak slightly, resulting in an internal corrosion buildup that will eventually prevent the valve from operating.

Water Level Control and Low Water Fuel Cutoffs
These devices perform two separate functions, but are often combined into a single unit. This method is economical, providing both a water level control function and the safety feature of a low water fuel cutoff device. We recommend, however, that both steam and hot water boilers always have two separate devices — a primary and a secondary low water fuel cutoff. They should be attached to the boiler through separate openings to prevent a restriction in the connecting piping from disabling both devices. Many jurisdictions require two such devices on steam boilers.

Piping should be kept open and free of scale or sludge build-up at all times.
Properly installed piping will use "cross tees" so the piping can be easily cleaned and inspected. A simple indicator that trouble may be developing in piping connections may show up when the float chamber of the low water fuel cutoff is flushed out or drained. The water level should quickly return to normal in the gage glass when the drain valve is closed. A slow return is a good indication that the connecting piping to the boiler is being restricted.

The most common water level control and low water fuel cutoff devices consist of two main components, a float chamber and an electrical switch operated by a float in the float chamber. A malfunction in either will prevent the cutoff device from operating. Malfunctions in the float chamber are generally the result of neglect; tampering and age most often cause those in the switch and associated wiring.

Exam Prep Course

As the water level in the boiler drops, there is a corresponding drop in the float. When the float reaches a preset position, it activates an electrical switch that shuts off the burner. Sludge and sediment accumulate in the bottom of the float chamber, and, if not regularly flushed out, will build up preventing the float from dropping down to the shut off level. Note that flushing the float chamber should not be considered as a test of the low water cut-out.

Be Careful When Testing

Low water fuel cut offs should be checked periodically for proper operation during the period when the boiler is operating. Since this test requires lowering the boiler water to the minimum safe operating level, qualified personnel should use extreme caution. Never allow the water level to drop out of sight in the water gage glass. This test should be done daily for steam boilers operating at more than 15 psig and weekly for those operating at less than 15 psig. In addition, a slow drain test should be done semi-annually on steam boilers operating at more than 15 psig.

In addition to these periodic tests of the low water device, the float chamber on the water level control and/or the low water fuel cutoff should be thoroughly flushed to remove any accumulated sediment. At least once a year, water level controls and low water fuel cutoff devices should be disassembled, cleaned and checked. These devices are an important part of boiler safety. Unless you are thoroughly familiar with them, have an experienced technician perform this type of maintenance.

The electrical switches and wiring are generally quite reliable and require little ongoing maintenance. **At least once a year, the switches should be cleaned and any dust or dirt removed.** The covers should be kept tightly in place except when opened for cleaning. If used and maintained properly these switches are virtually trouble free. However, if abused they can be a prime cause of boiler accidents. During the annual cleaning the wiring should be examined for signs that insulation is cracking. All connections should be tight.

Don't Bypass the Switches

It is not unusual for a maintenance worker to remove the cover and install a "jumper" wire to prevent the switch from operating. This starts out as a temporary convenience, often to "fix" a boiler that keeps shutting off on low water while being operated at high demand or as a temporary means to test other circuits in the control system.

This bypass can easily become a permanent and dangerous condition. A boiler that regularly shuts down indicates a very serious problem that could lead to a catastrophic accident. A jumper wire should never be permanently installed in a low water device. Only a qualified technician should use a jumper to test another circuit.

The Fuel System

The fuel system, particularly the burner, requires periodic cleaning and routine maintenance. Failure to maintain the equipment in good working order could result in higher fuel costs, the loss of heat transfer or even a furnace explosion. Modern fuel systems are very complex assemblies, consisting of both electronic and mechanical components. Over a period of time many things may go wrong — ignition transformers deteriorate or fail, ignition electrodes burn and become coated, fuel strainers and burner equipment become clogged, fuel valves become dirty and leak, air/fuel ratios drift out of adjustment, flame scanners become dirty. Many users wisely contract with their gas company or oil service company to periodically check and maintain their burner equipment.

Properly maintained equipment should be safe and reliable, but devices installed to assure safe operation are sometimes viewed as an inconvenience. The personnel who operate the boiler may tamper with or adjust these devices, thereby compromising operation of the boiler.

The safety feature most often adjusted is the burner purge cycle, designed to prevent furnace explosions caused by a buildup of unburned fuel in the furnace chamber. The cycle length is determined by the equipment manufacturer to purge fuel from a leaking fuel valve or an unsuccessful ignition sequence. It is annoying to have a boiler fail to ignite and then wait for the burner to go through another complete purge cycle. You may be tempted to shorten or even bypass the cycle. Don't! Doing so greatly increases the chances of a serious furnace explosion.

Exam Prep Course

The Water Gage Glass

The importance of proper cleaning and maintenance of the water gage glass, or sight glass, cannot be stressed enough.

The water gage glass on a steam boiler enables the operator to visually observe and verify the actual water level in the boiler. If not properly cleaned and maintained, however, a gage glass can seem to show there is sufficient water, when the boiler is actually operating in a low water condition. A stain or coating can develop on the inside of the glass where it is in contact with boiling water. After a time, this stain gives the appearance of water in the boiler, especially when the glass is completely full or empty of water.

Another problem that can be the indirect cause of accidents is for the connection lines to the gage glass to become clogged and show normal water levels when water may be low. The piping connecting the gage glass to the boiler should be cleaned and inspected regularly to assure it remains clear.

One final problem should be mentioned. Often, a boiler is operated with the isolation valves to the gage glass closed because the glass has been broken, or is leaking. Take the time to replace the glass, even if the boiler must be shut down. That inconvenience is nothing compared to the damage that may result from operating a boiler without a gage glass. Some operators routinely replace the glass and seals during annual maintenance, because it is so important to verify the actual water level.

The Stack Temperature Gage

A stack temperature gage is normally installed on a boiler to indicate the temperature of the flue gas leaving the boiler. A high stack temperature indicates that the tubes may be getting a buildup of soot or scale. Also, the baffling inside the boiler may have deteriorated or burned through, allowing gases to bypass heat transfer surfaces in the boiler. These conditions generally develop slowly over a long period of time, slow enough so the person who operates the boiler can become accustomed to the gradually rising temperature. Approximately 1 percent in boiler thermal efficiency is lost for a 40-degree F increase in stack temperature.

The Best Boiler Operator

Boiler Logs Are Important

The majority of boiler accidents can be prevented. One of the most effective tools is the proper use of operating and maintenance logs. **Boiler logs are the best method to assure a boiler is receiving the required attention and provide a continuous record of the boiler's operation, maintenance and testing.**
Because a boiler's operating conditions change slowly over time, a log is the best way to detect significant changes that may otherwise go unnoticed.
If a boiler is to be kept in good operating condition, someone who tends to the boiler must be responsible for its operation and maintenance. This person should have a good understanding of boiler operation and safety devices. Maintenance and testing should be performed and recorded in the log on a regularly scheduled basis. The responsible individual should initial the log to verify each operation performed, who performed it, and when it was done.

Disclaimer statement:

All recommendations are general guidelines and are not intended to be exhaustive or complete, nor are they designed to replace information or instructions from the manufacturer of your equipment. Contact your equipment service representative or manufacturer with specific questions.

Exam Prep Course
Chapter 9: Boiler Shut-Down

During normal operations, the boiler water chemistry is carefully controlled so that the dissolved/suspended material is conditioned to prevent hard deposits on boiler metal. These dissolved/suspended solids are maintained in suspension by water circulation and the action of the treatment chemicals. When a boiler is shut down or drained, this material (sludge) may settle and bake on tube surfaces; it may become so adherent that mechanical (turbining) chemical cleaning may be required. At worst, there are large piles of sludge in the mud drum and in the lower tube ends which cause the customer and/or boiler inspector to feel that the deposits developed during operation, and thus unjustly criticize the treatment program. Outlined below is a procedure that if followed will minimize the total amount of sludge left behind when a boiler is opened.

Shut-down Procedure

1. Three to five days before a scheduled shut down, increase the blow down by 50%.

- If possible, increase the alkalinity to at least 500 ppm. Go as high as possible without causing foaming or carryover.
- Due to the increased blowdown rate, the feed rate of the scale inhibitor and oxygen scavenger must be increased so as to maintain the normal boiler water residuals.
- If possible, increase the sludge conditioner level in the boiler water by 50 to 100%.

2. During the last twenty-four hours before shut down, decrease the continuous blowdown and increase the manual blowdown.

- Frequent short bottom blows are better than fewer

longer blows.
- Generally it is sufficient to hold each mud drum blowdown valve open for about 5-10 seconds every one to two hours.
- Once the load is dropped from the boiler, include the header blowdowns as part of the manual blowdown procedure.

3. When the load is dropped from the boiler, continue bottom blowdowns until boiler is cool and safe to work on.

4. As soon as possible after the boiler is opened, wash down the boiler watersides, preferably with soft water.

Exam Prep Course
Chapter 10: Boiler Inspection

It's extremely important that your boiler system maintained, adjusted, and inspected annually to ensure optimum performance, efficiency and safety. An annual inspection allows a technician to identify and address issues before they cause more damage to your system.
High Efficiency Boilers, and Hydronic systems are expensive to install but highly efficient if installed and maintained properly.

Maintenance checklist

We offer annual maintenance service for a wide range of hydronic systems including indirect water heaters and in-floor heating.
Here we'll perform the following tasks, depending on the unit:

Boiler Service Checklist

- Code requirements and new updates are considered during boiler inspection.
- Sensors are checked and cleaned, if required.
- Gas pressures are checked and adjusted if necessary.
- Glycol in-floor heating freeze points are measured and adjusted (if applicable).
- Boiler burner chamber is inspected and cleaned (or flushed) as needed.
- Controls (thermostat), safety, auxiliary controls, wiring and connections are all checked.
- Amperage draws and performance of electrical parts are tested to ensure proper performance.
- A Combustion Efficiency test is taken and printed, adjustments are made to insure proper manufacturer set points.
- A Carbon Monoxide (CO) test is also conducted to ensure the fuel is burning safely within your home.

The Best Boiler Operator

- All circulators are checked and lubricated if applicable.
- General water piping, pressures, and expansion is checked and adjusted if required.
- A pH test is done to determine water qualities within system.
- When completed, we'll give you a comprehensive written report of the details of the service. The technician will make any recommendations at this time, and answer questions about your system

Which do boilers require an annual inspection?

Ans. Inspections required on power boilers (over 15 psi steam or vapor pressure): An annual internal certificate inspection and an external inspection while under operating conditions (approximately six months apart).

What is the purpose of annual inspections?

Ans. Inspections are used for the purpose of determining if a body is complying with regulations. The inspector examines the criteria and talks with involved individuals. A report and evaluation follows such visits.

Exam Prep Course

How to prepare for a boiler inspection?

1. Shut down the boiler using proper shut down procedures as required by your boiler operating procedure.
2. Lockout and tag all steam, water, and fuel valves, the ignition system, and electrical disconnects.
3. Allow the boiler to cool completely, 24 to 48 hours depending on the style and size of the boiler.
4. Open manholes and handholes to drain and flush the boiler
5. Clean out the combustion space. If firetube boiler punch tubes.
6. Inspect the water side to make sure all scale and sediment has been removed from the water side.

The Best Boiler Operator

7. Inspect combustion chamber and check refractory for cracks or direct impingement of the flame.
8. Wick the water side to assist in drying the heating surfaces
9. Call for inspection and have boiler certificate ready for signing.

Exam Prep Course
Chapter ii: Boiler testing

Boilers can be tested in many ways. Some tests are done to test the capacity and others are done to test the operation of a component. All are important to the safe and efficient operation of the boiler.

What is an Accumulation Test?

To detect the safety valve is suitable for this boiler or not. To limit the rise in boiler pressure under full fire condition.

This test is carried out a new boiler or new safety valve.

1. Shut off feed water
2. Closed main steam stop valve.
3. Increase cut off pressure of boiler.
4. Bypass high pressure cut off of the boiler.
5. Arrange the boiler fire rate to a maximum.
6. Safety valve will be lifted during the test.
7. The test is carried out as long as the water permits in the boiler.
8. Accumulation pressure should not exceed 10% of the working pressure in the specified time.
9. Specified time is 15 mins for a smoke tube boiler and 7 mins for water tube boiler.

Setting of safety valves:

1. Take standard pressure gauge (approved by surveyor) for accuracy.
2. Fill up water up to ¼ of gauge glass level, and shut the main steam stop valve, feed check valve.
3. Without compression rings, hoods and easing gears, reassembled the safety valves with spring compression less than previous setting.
4. Raise the boiler pressure to desired blow off pressure.
5. Screw-down spring compression nuts of any lifting

THE BEST BOILER OPERATOR

valves, until all are quiet.
6. Arrange to have the desired steam pressure
7. Adjust each valve in turn: Slacken compression nut until the valve lifts. Screw-down compression nut sufficiently enough, so that when the valve spindle is lightly tapped, valve return to its seat and remain seated. Measure gap between compression nut and spring casing. Make a compression ring equal to this gap, and insert under compression nut. Gag the spindle of this safety valve, to prevent opening, while remaining valve is being set.
8. Remaining valve is again set and insert compression ring.
9. Remove gag and retest both valve to lift and close together.
10. Cap, cotter and easing gear to be refitted
11. Caps and cotter pins padlocked to prevent accidentally altering the setting.
12. When the surveyor satisfied the setting pressure, easing gear should be tested.
13. All safety valves set to lift at not greater than 3% above approved working pressure (design pressure).

What is an Evaporation Test?

Low-water cutoff, **evaporation test (steam boiler)**– While the **boiler** is in operation, shut off the feedwater pump and monitor the **boiler** water-level. The low-water cutoff should shut down the burner before the water level goes.

What is a Hydrostatic Test?

A procedure that employs **water under pressure**, in a new boiler before use or in old equipment after major alterations and repairs, to test the boiler's ability to withstand about 1½ times the design pressure.

What is a Flame Cutout Test?

Exam Prep Course

Covering the flame sensing mechanism so that the system shuts off the flow of fuel to the furnace. Maintaining a good flame in a boiler is the key to a good boiler performance. Unstable flame is always a threat for boiler furnace explosion, which can lead to a large outage of the boiler and economic loss. When a flame failure occurs in a boiler, the boiler desk operator and the local operators will have to act immediately and bring the boiler back on line with all safety taken into consideration. It is always seen that the greatest number of explosions in boilers takes place during start-up and shut-down. It is during this period that the probability for unburned to accumulate in flue gas path of the boiler is very high. Hence it has become a practice of all boiler designers to interlock purging the boiler with boiler start-up. Flame failure in a boiler can be due to many reasons.

The specific causes

- Heavy oil or warm-up oil trip valve closing
- Sudden reduction in mill feeder speed to a minimum
- Loss of ignition energy
- Flame scanner trip due to poor and unstable flame
- Malfunctioning of flame scanner
- Scanner air fan trip and slag build-up
- Sudden coal hang-up in one or more mills
- Coal feeder trip
- Electrical supply failure
- Air flow below 30% of total air flow
- Tripping of all forced draft air fan
- Tripping of all induced draft air fan
- Very high furnace pressure

These causes lead to flame failure individually or in a few a combination. This will depend upon when the flame quality becomes unacceptable.

The plant response

The Best Boiler Operator

- In many cases it is possible to see fluctuation in furnace pressure.

Flickering of the flame scanner can be seen when flame quality deteriorates Tripping of the primary air fan or very low primary air header pressure is more probable to occur just before flame failure.

After flame failure immediately the unit trips

The boiler desk operator immediate action

- The boiler desk operator will have to start boiler purge
- Assess the reason for flame failure
- Light-up the boiler after the purge is complete and local operator gives clearance

Local operator immediate action

- Check furnace and report to boiler control room
- Check flame scanners for any possible slag accumulation
- Check mills and restore them for starting
- Check oil trip valves for proper function
- Make sure oil guns are in position for light up of boiler

Exam Prep Course
Conclusion:

As you have completed this book it becomes imperative that you go through it a few times. The book is designed to help its readers to pass municipal license examinations and give a layman's explanation to boiler operator applications. No book on the market will prepare a student if they do not want to commit to studying. However, this book sets itself aside from the rest because it removes the filler and provides only the substance needed to have you better articulate and communicate the basic principles of boiler operation.

You should spend no less than ten hours a week studying this manual. If the simple language provided here remains an obstacle then the reader should attempt to place the subject matter in its own language following the format outlined here.

After 30 days of studying schedule your examination. Often times readers will study and study but never schedule an examination. Then life happens and they never go test. After so long passes the skills dull and they need a refresher all over again. Well this book is that refresher. The author also suggests purchasing the audio version of this book to listen as you go about your day to help commit its subject content to memory and in a way that the listener will be able to regurgitate its content.

Prepare, commit, train, apply and repeat! That is the key to you getting you passing your next license examination. Good luck!

www.ingramcontent.com/pod-product-compliance
Lightning Source LLC
Chambersburg PA
CBHW022116170526
45157CB00004B/1668